Praise for
From the Garage to Mars: Memoir of a Space Entrepreneur

Scott's story of building a company from an invention of hardware-store parts and wax, to motors on Mars, takes the reader along for a wild ride. It will inspire and inform space aficionados, entrepreneurs, and business leaders, alike, and is a testament to our country's entrepreneurial spirit.
—Andy Weir, Best-selling author of *The Martian, The Hail Mary Project*

An amazing and FUN story of Scott Tibbitts, his company Starsys, and their journey in support of our space exploration programs. Great management tips and insight into the thinking of an entrepreneur – I finally understand what makes an entrepreneur tick!
—Jim Voss, US Astronaut, 6 flights, co-holder of world record for longest spacewalk

"From the Garage to Mars" is a fascinating story of how a few men with a simple idea founded a company that was instrumental in the success of the Mars rovers, Spirit and Opportunity.
—Pete Theisinger, Project Manager, Mars Exploration Rover Program

In "From the Garage to Mars," Scott describes the complexity of being an integral part of space flight, which at the same time, how those complexities reinforce leadership principles that can, and should, be relevant to any organization or leader. This book elevates the concept of leadership with an emphasis of not just leading others, but also those unique experiences that will assist you in leading yourself. The story of Starsys that Scott describes will give you an appreciation of the company's desire to push its limits, and as a leader, this book will help you push yours.
—Jeff Potter, Founder and CEO of Manifest; former CEO of Frontier Airlines

"From the Garage to Mars" manages something rare—to bring the reader along for the ride of punch-your-fist-into-the-air highs, and the soul-wrenching lows that are a part of every entrepreneur's journey. The reader shares the jubilations and heartaches as Scott goes from hardware-store invention to spaceman. A must-read for entrepreneurs excited by space travel, as well as space exploration enthusiasts fascinated by entrepreneurship.
—Dave Cohen, Founder and Chairman of the Board: Techstars Entrepreneurial Incubator

This is a terrific book that brings us inside the mind and heart of what it is like to be an entrepreneur in a highly competitive and technical industry. Throughout these fascinating pages emerges the unvarnished story of the successes, failures and decisions behind a true pioneer of the commercial space industry. It is a must read for anyone interested in space or in doing what Scott has done.
—Mark N. Sirangelo former President, Sierra Nevada Space Systems, Entrepreneur in Residence, University of Colorado

This is a lively and engaging story. Scott Tibbitts reminds us that that entrepreneurship is not so much a technical process, it is a human process. What separates entrepreneurs from inventors is that entrepreneurs create organizations. This is a wonderful story of the creation of an organization.
—Andrew Aldrin, President, Aldrin Family Foundation

I loved "From the Garage to Mars." Scott's personal journey is inspiring, heartbreaking, laugh-out-loud funny, and entertaining. His approach to business, and in particular his ability to use corporate culture to transform business performance, is inspiring.

In the midst of a world of work where corporate fun, caring, and wellbeing are eroding, Scott shares a vulnerable, honest account of his successes and travails that show that caring and fun can not only coexist in a high-performance organization, they are necessity. In addition, in the telling of his story, Scott shares dozens of tools and processes that will be light-bulb moments for business leaders looking to transform THEIR companies. ... "From the Garage to Mars" brings emotion and real-life hacks to one wonderful must-read story!
—Chris Cummings, Group CEO, Wellbeing at Work

Scott Tibbitts' narrative is a beacon for all dreamers looking skyward, proving that the vastness of space is no match for the boundless ambition of a driven entrepreneur. This book is not just about reaching Mars; it's about the journey of overcoming earthly challenges with a celestial vision, encapsulating the essence of true entrepreneurship.
—Thomas Frey: Best-selling author, CEO of Futurati

Scott is an exceptional storyteller, space pioneer and so much more. He has shared deep wisdom in this book that is a must-read for budding entrepreneurs and space enthusiasts. With this highly inspiring book, I felt the spark to get back deeper into my passion of space. Prioritizing Wellbeing (with Quality of Life and Work-Life balance) and with his 2Fs - Fun and Family (or Human Connection) go a long way in aligning with one's Ikigai or MTP (Massively Transforming Purpose). Curiosity, Imagination, Passion to Obsession, Dedication, Perseverance are key traits of Innovation that Scott has explored well. The tips shared are nuggets of immense value.
—Anuraj Gambhir, Global Expert Faculty, Curation Partner, Singularity University, Australia

"From the Garage to Mars" was a hugely enjoyable read and gives unparalleled insight into the journey businesses take to transition from a start-up to a world-class industry leader. Scott's humor and candidness shines throughout when exploring the growing pains of organizational and cultural change. Highly recommended for those who have an interest in space exploration, engineering, or business excellence.
—Richard Nelson, Airbiotics Talent

Scott Tibbitts is a rarity. An exceptional leader and storyteller who knows what it takes to make the human experience of business work. If you are looking for an in-the-trenches experience of what the entrepreneurial ride can be, as well as a guidebook to how fun can create a legendary and aligned corporate culture, this is it.
—Edgar Papke, Leadership and Business Psychologist, best-selling author

Embark on a journey from humble beginnings to interstellar success with CEO Scott as your guide in "From the Garage to Mars." This book not only makes for an engaging read but also provides actionable strategies for turning your entrepreneurial venture into a dynamic and enjoyable company. Written by CEO Scott Tibbitts, known for his transformative leadership, this book shares invaluable tips on creating a vibrant company culture, designing an innovative workspace, and fostering collaboration.

Whether you're a seasoned CEO, an aspiring business leader, or just starting out, Scott's insights will empower you to inspire your team, drive innovation, and successfully navigate change.
—Stephan Reckie, Executive Director of the Global Entrepreneurship Network for Space

"From the Garage to Mars" is a story of hard work and perseverance, but at its heart are family, belief, and the people who made this incredible journey possible. It's a story of opportunities won and lost from an entrepreneur who has experienced it all. Scott Tibbitts shows us the all-too-human side of entrepreneurship, the genius, and the folly of aiming for the transformational, the exhilaration and the heartache of dealing with its consequences. Just as he has in mentoring our young space entrepreneurs in Adelaide, South Australia, Scott Tibbitts shows us that this journey to business success is one about supporting the people around you as much as it is about creativity, technical brilliance, and business savvy.
—Ben Rowley, Program Manager ICC Innovation and Collaboration Centre, University of Southern Australia

Scott Tibbitts tells a riveting true story about the ups and downs of a simple device and a mantra of "fun" that became a spacecraft mechanisms company with over 3,500 devices flown in space and zero failures. A must-read for anyone interested in entrepreneurship, space, and the human side of business.
—Bert Vermeulen, Professor of Entrepreneurship,
Colorado State University

When I read "Memoirs of a Space Entrepreneur," a dream-fulfilling message returned. The creativity, intelligence, and tenacity of the author were the bases of his success, which he describes compellingly in this book. Scott's story is both heartwarming and optimistic.

Apart from the sheer fun one gets from reading that book, there is an extra bonus: a recipe on how to build a devoted team that is ready to follow their leader under any circumstances. Enjoy the reading!
—Tom Zawistowski, PhD, Space Scientist extraordinaire

If you want to know what it is like to lead a small company pushing the frontier of space technology, this is the book to read.
—Dr. Robert Zubrin, President of The Mars Society,
Author of *The Case for Mars* and *First Landing*

FROM THE GARAGE TO MARS

Memoir of a Space Entrepreneur

FROM THE GARAGE TO MARS

Memoir of a Space Entrepreneur

SCOTT TIBBITTS

Milwaukee, Wisconsin

Copyright © 2024 by Scott Tibbitts
All rights reserved.

Photographs from the author's personal archives
unless otherwise noted.
NASA photos are in the public domain.

Published by HenschelHAUS Publishing, Inc.
Milwaukee, Wisconsin
www.henschelHAUSbooks.com

ISBN: 978159598-987-1
LCCN: 2024934839

Printed in the United States.
Second printing.

TABLE OF CONTENTS

Preface i
Prologue 1
Something from Nothing 7
Bona Fide 15
Wax in Space 25
Choosing the DNA 33
More Than Wax 45
Becoming a Space Motor Company 59
Serendipity / Nudges 63
On Mars 69
Losing Kurt 85
The Morning Meeting 95
Graffiti 125
Lunch with Cary 135
Throwing Hats Over Fences 143
Entrepreneurial Hell 151
A Million Dollars in Ten Days 165
SpaceDev 171
Decision 183
Founder in an Acquired Company 193
The Red Button 201
True North 215
Epilogue 229

Acknowledgments 237
About the Author 239

PREFACE

IT WAS NOVEMBER 2015. As respite from an early Colorado snowstorm, my daughter Alyssa and I were at the Regal Theater in Boulder to see a movie I had been looking forward to watching for years.

I was a space nut who had somehow leveraged an engineering degree to become a part of NASA's exploration of space. Twenty-five years earlier, I had invented a new kind of spacecraft motor made of wax and hardware store parts that ultimately led to my company providing thousands of space motors for hundreds of spacecraft, including a dozen spacecraft that had explored Mars, each motor working perfectly.

I had followed in the footsteps of my father, Ted Tibbitts, a professor of horticulture at the University of Wisconsin who had followed his passion for space by becoming a world expert in space crops. Dad was responsible for the first-ever crops grown in space on Columbia Space Shuttle flight STS 73 in 1995—five small, red Norland potatoes.

Alyssa and I were there to see *The Martian*, written by Andy Weir, directed by Ridley Scott and starring Matt Damon. I was tingly with anticipation because of two connections to the movie: first, to the author, Andy, who like me, had been a nerdy science fiction nut, loved Heinlein and Asimov, and harbored the crazy idea that he might also be a writer.

And second, because the signatures of our family—my wife Jackie, son Ryan, daughter Alyssa, and I—were on the north pole of the Red Planet, carried there by the Mars Phoenix lander.

FROM THE GARAGE TO MARS

The movie started. From the onset, it was a remarkable, spot-on representation of how NASA balances the competing forces of exploration, politics, technology, and tragedy. The story starts with Matt Damon's character, botanist Dr. Mark Watney, on Mars with a group of fellow astronauts. A slight thrumming of familiarity coursed through me. *Isn't a horticulturist a kind of botanist?* Dr. Watney then becomes stranded on the planet. His survival depends on a crop of potatoes that he tenderly cares for, the plants keeping him alive.

I startled. That was my dad's research. That was the crop that Dad had talked about incessantly for 20 years. That was the vegetable I had seen launched into space, in person, from Kennedy Space Center in 2005. I choked up, flooded with parental pride. My dad, the guy who had taught me to fish, and on who's back I rode around the living room, had saved Matt Damon's life.

The story continues. The Mars-base plant growth module was ruined by a propellant explosion. Dr. Watney is brought to tears by the loss of his plants. His chance for rescue was now a long drive across the planet to the Pathfinder spacecraft, where he digs out the Rover, reprograms the computer, and uses motors at the top of the spacecraft's mast to point at various signs. Ultimately, he is able to communicate to Earth that he is alive, which then leads to his rescue.

Those are the motors that my company, Starsys Research, had built, and were still working 47 movie-years later.

Dad had saved Matt Damon's life on one side of Mars, I had saved his life on the other! The apple hadn't fallen far from the tree.

Four years later, my dad was turning 90 and family members from all over the country were flying in for a birthday party in Texas. I reached out to Andy Weir, and to my delight, he responded. I told him the story and shared an ask. If I FedEx'ed a potato and a print copy of *The Martian*, would he sign the book with a

PREFACE

personal message, take a picture of him, the book, and the potato, and send the picture and the book back? It was too strange a request for him to pass up, and a couple of days later, the book and the image arrived. I put the picture, as well as several others from Dad's space potato adventures, into a collage, and gave both to him on his 90th surrounded by more than 20 of his family.

This would have seemed a miraculous coincidence if it hadn't been for the dozens of miraculous coincidences that had occurred over the 30 prior years that led to that moment.

This is that story.

Andy Weir, author of *The Martian*

PROLOGUE

IT WAS LATE JUNE 1997. As the founder and CEO of Starsys Research Corporation, for years, I had occasionally reminded my wife, Jackie, that someday we could be famous for the wrong reasons, imagining some monumental space disaster and an accompanying news headline:

> **"Space Motor Made by Small Space Company in Boulder, Colorado Fails. Billion Dollar NASA Mission Lost. CEO/Entrepreneur Scott Tibbitts says, "I'm stumped ... it seemed to be working just fine before we put it on the rocket."**

The comment was attended by a nervous chuckle from us both, being closer to the truth than was comfortable. Our company of 75 wet-behind-the-ears rocket scientists was making the things that opened telescope covers, released solar panels, and separated spacecraft from launch vehicles. If a latch on your car gets stuck, you call a mechanic. If a latch gets stuck in space, Congress wonders if they should continue to fund NASA.

Spacecraft failures are rich with irony with billions of dollars lost from the occasional "Doh!" moment. With the ridiculous complexity of space travel and the trying for the nearly impossible, it's the simple things that are missed. You might think that the explanation for the big failures might be something like "Our Fourier transform analysis failed to take into account secondary shock waves of the plasma thrusters." Nope. More "Whoops ... what *were* we thinking?"

FROM THE GARAGE TO MARS

Consider these few incidents: The Mars Lander that augured into the Red Planet, its landing rockets turning off 100 feet too high because a confused switch thought it had already landed. Or the Earthview spacecraft, rechristened Seaview, after plunging into the Pacific when the latch to release the spacecraft from the rocket simply refused to let go. Or the Genesis spacecraft that plowed into the New Mexico desert floor at 500 mph because a parachute release latch hadn't been plugged in quite right.

Space mechanisms are often to blame for spacecraft failures, and that was our chosen business. Added complexity was brought to the mix by the unusual way our motors worked: rather than use magnets and wires, ours were powered by melting wax ... our secret sauce and the reason our group of wanna-be-rocket-scientists was in this business.

The nature of the space biz is such that the CEO provides his or her phone number so as to be immediately available in the case of unforeseen "anomalies"—space-speak for "WTF? What went wrong? Who built this thing?" events that only occur at ungodly hours. The phone rang at 1:15 a.m. that day in late June. It was one of the Chandra Telescope program managers on the line, and the call I hoped wouldn't happen was happening. Chandra was the next of NASA's great observatories: a series of billion-dollar spacecraft designed to see things we have never seen before. The first of these was Hubble, looking at stars with visible light. Chandra would do the same but with X-rays discovering things billions of years old, never before imagined.

The call was brief and the words sobering: "There was a problem in test this evening. The telescope cover did not open. We believe your motor is likely the cause. The launch schedule is now paused." and an understated, rhetorical question "Could you be available for a telecon tomorrow morning?"

PROLOGUE

NASA had decided to send the motor back to us for analysis. They were shipping it that morning, and they asked if they could help with the examination. Trial lawyers have a fundamental tenet: "Never ask a question you don't already know the answer to." Space gizmo builders have the equivalent: "Never open a failed device for the first time with the customer looking over your shoulder." Too often the desire to quickly find the smoking gun can lead to the misperception that anything out of the ordinary is likely the cause of the failure. If you are going to show your dirty laundry, look through the pile ahead of time to be able to explain the stains.

But at that moment, in a temporary flash of self-righteous hubris, I responded that any customer observers were welcome. I said that we were sure of our design, and we welcomed them joining us as we opened the mechanism to look inside. Two days later, as I drove into our parking lot at 7 a.m., I recognized my pride-driven invitation had backfired. The lot was overflowing with white rental cars, belonging to a dozen high-level NASA representatives and customer program managers from both coasts there to watch firsthand.

As I walked into the building, I could feel the buzz of the group and it was not good. I realized they now understood that a critical part of the mission had been built by a company of tattooed engineers and technicians; a company with a large sign in its reception area that declared "FUN" as one of its guiding principles. Our visitors were used to seeing impressive buildings with shiny, brightly lit assembly areas and viewing balconies that looked like an operating room. Instead, they were here, crowded into a small room in an industrial park in Boulder, Colorado, next door to an elementary school, standing among space technicians ten years too young, sporting full-sleeve tattoos.

FROM THE GARAGE TO MARS

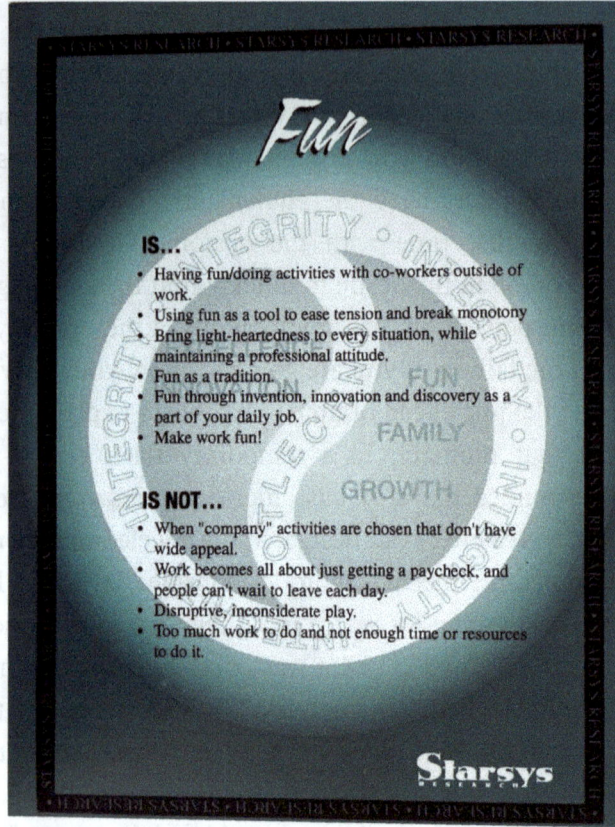

Starsys "FUN" Value Poster

I had made a monumental mistake with the open invitation. Millions of dollars had been lost over the past couple days from the halt in testing. The mechanism, which had been buttoned up a half year ago by a technician, was about to be opened and subjected to the extraordinary scrutiny of a group of tough, exacting space industry veterans. I remember wondering if the technician, who had likely been working late at night finishing his work on the mechanism, had been thinking at the time about getting it just right before going home, or had he been thinking about meeting his friends for Marg's at the Rio Grande restaurant downtown?

PROLOGUE

The room was thick with tension, made worse by the inability of the air conditioning to keep up with the heat from the lights and the dozen or so visitors pressed into a small space. I could sense the conversations that were happening covertly. *I think we've found the root cause. A company like this has no business making flight hardware. Who decided that a motor driven by melting wax could be trusted for a billion-dollar spacecraft? How are they going to find a traditional replacement motor quickly?* I could feel their belief that they had found their smoking gun, as well as their foregone conclusion that we were a bunch of daring renegades who had ultimately failed.

I knew that anything out of the ordinary, such as a fingerprint, an unexpected metal shaving, or any other sign of damage would be immediately considered a root cause. This was worse than "Did our device fail?" This was Starsys opening its closets that might hold unexpected skeletons to an audience ready to banish us from the industry if they found a single flaw.

The mechanism sat peacefully on the workbench. It was a brick-sized box with an aluminum surface that glistened like a piece of expensive jewelry. Standing over it was our chief of technology, Scott Christiansen, who had begun to open its cover while our guests looked intently over his shoulder.

One at a time, the six Allen-head fasteners on the top cover surrendered to Scott's wrench. Despite the tension in the room, there was no rush and Scott took his time, with each removed screw slightly sharpening the room's angst. As the observers leaned in to see the cover come off, I was squeezed out ... perversely reminded of an old Warner Brothers cartoon in which dozens of missiles, explosives, and bazookas surrounded a road runner pecking away innocently at a pile of bird seed.

In that moment, I realized that the future of my family and the 75 people in the company was dependent on what was underneath that cover. This had become a *Very Bad Idea*. Inside the box was the result of a ten-year experiment on how to run a space business differently, with such heretical fundamental principles as Let's Have Fun, Work-Life Balance Matters, and We Are Family; an experiment in which all of us would occasionally say, "Can you believe we get to do this???" I pushed my way between the bodies in front of me to watch as my future was revealed as if looking into a crystal ball.

It was a pivotal moment for our company. However, to truly understand the moment, we must go back to the beginning … to an invention made from $7.25 in hardware-store parts that began a journey to the planets.

SOMETHING FROM NOTHING

INVENTIONS THAT BECOME entrepreneurial ventures are birthed from one of two mothers: 1) The solution to big problems—getting taxis (Uber), splitting restaurant bills (Venmo), tying shoes (Velcro). And 2) things we didn't know we needed—iPhones, Snugglies, and Air Fryers.

But occasionally, inventions are born a bastard child, fatherless, no immediate application, but so compelling they create a question that won't let go; "There has got to be a use for this." Teflon, Post-Its, and Silly Putty came to be this way.

Starsys Research was created from such an invention. A device that created a visceral and immediate response of "That is super cool! There has got to be a use for that!" We did not expect the answer would lead us down a path of becoming rocket scientists.

The invention was simple and was brought to life in its original form by my partner Daryl Maus within his company Maus Technologies. I had met Daryl a couple of years prior, when he introduced himself as an entrepreneur, the first person I had met that had the *cajones* to introduce himself as someone who created companies for a living. I was fascinated, and we quickly became good friends, circling each other as we tried to find ways to work together. A couple of years later, he was starting a new company based on a home water heater he had invented and convinced a US company to market.

I was intrigued enough to quit my job as a nuclear weapons engineer and become the Director of Engineering for our three-person, one-dog company; 1,000 square feet of Class C office space in an industrial park, with a roll-up garage as the back room, and a live-in homeless genius, deep into the Asperger's spectrum, who slept all day and read physics textbooks all night, trading janitorial duties for the ability to live in the back half of our "R&D lab." As well as a renegade, PTSD-afflicted Vietnam-vet-machinist who had changed his name to John Galt, who quietly hummed "If I Had A Rocket Launcher" without irony while he worked, and who was wanted by the police in Boulder County for orchestrating a Humane Society jailbreak, releasing all the dogs shortly after the police incarcerated his license-less mongrel companion.

Daryl's water heater prevented someone from taking a cold shower once the hot water ran out. Doing that required a small piston to open a valve at a particular temperature. The solution was a small device, wonderful in its simplicity: a small cylinder, filled with paraffin wax, a seal, and a piston.

In addition to being great for canning and lining Starbucks cups, paraffin wax has the peculiar characteristic that it expands when it melts. A lot. It turns out that it expands more than pretty much any other material on the planet. This is a pain in the ass for the manufacturers of wax lips and candles, but a peculiarity can be a feature. Melt wax and it expands more than 15%. Put the wax in a small cylinder where it's prevented from expanding and you get thousands of pounds of pressure as it tries in vain to expand. Put a seal around a piston at the end of the cylinder and all that pressure can lift hundreds of pounds over a distance of an inch or two. This wax actuator was the key to the new water heater: when the water reached a certain temperature, the piston would open a valve, when the water cooled, the valve would close.

What interested me most was the simplicity and "wow" of the little device. We made the first versions from copper pipe and seals bought at the hardware store for $7.00. The initial experiment was about the size of a tube of lipstick with a small nipple of a shaft sticking out of one end. When you put it in hot water, the piston extended an inch with enough force to lift 50 pounds. A thing of wonder to an engineering geek. Using the heat from a cup of coffee to lift the equivalent of a small child.

The water heater business was cooling off, and my wife Jackie had tired of my complaints over dinner that I was not cut out for designing a household appliance. She finally said what she had been thinking for months. "Instead of complaining, is there something you can do about it?".

The next day, I went to Daryl with a proposition.

"How about you pay me half-time, I look for a new use for the wax actuator, and if I figure it out, we'll be 50/50 partners in whatever comes?"

After a couple-second pause, Daryl said "Deal." We shook hands and that was it. He went back to his computer, and I was let loose to create my own company.

I've come to realize how often life changes course in an instant. Your wife asks, "What are you doing to do about it?" A partner says, "Deal." This was to be a recurring theme for the next couple of decades.

What drives entrepreneurs is an unrealistic, unshakeable optimism that they have in their hands something that someone will want, that will change the world, and no matter how many people say, "I'm not interested," that's only because they don't share the Truth that the invention is worthy. It's hard not to admire the compulsive disorder in someone who can hang up after hearing, "I'm not interested. We have no use for that" for the 50th time, and ten seconds later, be psyched for the next call, sure that the 51st will recognize what is so plainly obvious.

I spent months making phone calls to various high-tech industries; medical (safely and slowly inject insulin!), solar (point solar collectors without power!), nuclear (move control rods in a harsh environment!), and ordnance disposal (remote control wire cutters!).

Then I cold-called NASA.

You don't actually just call NASA. You scour phone directories and organizational charts to get you close enough to someone who might be interested enough in your story to connect you to someone who might be interested in your device.

My calls were surprisingly welcomed. The earnestness of an inventor calling and saying, "I've got this really cool thing you might be able to use. Want to hear about it?" is an echo of a kid asking a parent, "Wanna hear a neat riddle?" Add that to a NASA full of smart people who love invention of any kind. Ex AV, chess, and physics club members for which the sting of rejection for their nerd-ness remains, and you get engineers who will go out of their way to help others they recognize from the same tribe.

My starting place was NASA's Jet Propulsion Laboratory, the NASA center that made the spacecraft that went to the planets. It took two calls to reach Carl Marchetto, head of a group that built spacecraft instruments. After about ten minutes of explaining how a wax actuator might be able to help cool spacecraft, he casually said something that still gives me goosebumps when I think back: "Sounds interesting. Why don't you come out and show it to us?"

I ran up my credit card a bit further and bought a $250 ticket to Los Angeles. I packed up a

---

Tibbitts Tip
Everyone is Approachable...

It is possible to connect (not just reach, but connect) with pretty much anyone *if* you have something that you are sure is of value to them. Cold-calling NASA was simply a matter of looking through the directory to find the person responsible for the area of NASA I thought I could help and reaching out to them. A key of Starsys' success, distilled down: Find the people with the problem we can solve, and connect. The same is true with running into big challenges in your life or business. There is always (almost) someone who can help with the solution. Find them and ask for their help.

couple of the $7.00 actuators in a small cardboard box. Daryl suggested I include one with a heater inside and call it a "linear motor." I was sure that was a waste of time but humored him and put one together. It was a prescient suggestion. I headed to Pasadena.

I vividly remember the look, feel, and smell of the Jet Propulsion Laboratory (JPL). Set against the foothills of Pasadena, with eucalyptus trees providing a faint scent of steam rooms, JPL feels more like a college campus than a NASA center. Once through the gate, I was walking shoulder to shoulder with the engineers who had put landers on Mars, and sent spacecraft to Jupiter. Models of the spacecraft I had grown up with were sprinkled throughout the campus. As a courtesy, I was taken to see the Galileo spacecraft, being readied to go to Jupiter. I could not believe I was here, in this world, being welcomed. I remember thinking how wonderful—but impossible—it would be to become one of them.

The meeting was in a conference room with sturdy gray furniture that looked like it had been purchased during the Cold War. The group made mechanisms for the spacecraft that were headed to the planets. Latches, motors, robots. Structures that unfolded like origami. Simply stated, they made the stuff on a spacecraft that moved. A device that lifted weight when it warmed was of little interest. But when they saw the "linear motor" that when powered, pushed mightily, they all leaned in.

John Kievett was the head of the group. Wise from the mistakes and successes of the past 30 years, he was long on curiosity and short on protocol.

"What happens if I double the voltage?"

The question was not rhetorical. Before I could grab his hand, he cranked the power supply to 100%. I was horrified, ready for a small puff of smoke and a thin spray of flaming wax. Engineers patting out small fires on their button-down shirts. A quick escort to the gate and 200 more phone calls to find someone similarly interested in the device. Instead, the actuator just pushed more weight faster.

"This thing is amazing. We could use these. Could you make these out of space-suitable materials?"

"Absolutely!" (Note to self: *What are space-suitable materials?*) "What would you use them for?"

John explained that for years, NASA had used explosive bolts to open covers from telescopes and to release antennas and solar panels from spacecraft. The problem was the explosives were like a shotgun shell going off inches from a $25M space camera. Making a space camera withstand that kind of shock made it a $26M space camera. And you could never actually test the one that you would use in space. The next best thing was to test 99 out of 100 to make sure the one that you sent into space would probably work. The running joke amongst the test engineers was, "Yup, that would have been another good one to use" as they worked through the 99.

John explained that the next big NASA mission would be called Cassini. The spacecraft would explore Saturn. Our wax devices could replace the explosives on the spacecraft. As they walked me to the gate, they gave me phone numbers and encouraged me to keep in touch. They had no budget for new products, but if I were willing to work on it, they would be willing to help me design a space version of our wax motor. They said if I built it, the rocket scientists would come.

Within a couple of hours, I was back in Colorado, driving up Highway 36 from the airport to home, replaying the events of the day. As I crested the hill and saw Boulder lit up and laid out in front me, I had a sudden intuitive flash. Something many years later I began to call a *Nudge*—a flush of gooseflesh followed by a singular thought.

SOMETHING FROM NOTHING

"That meeting might have changed my life." Followed immediately by more reasonable self-thoughts. "Things like that don't just happen. You don't go to NASA and say, 'Look at this neat thing. Can you use it on your rockets?' And then start a space company.

By the time I was home, the two had merged into one thought, "But maybe...."

(Courtesy of Wikimedia Commons)

BONA FIDE

THE FRAGILE NATURE OF A nascent entrepreneurial venture requires nurturing from a slowly growing group of believers who have been seduced into the same vision of what can be. These people are not only those who work within the venture, or invest, but more importantly, those who can influence its success, and who have no reason to do this other than a strong desire to help make it so. This is one reason why cranky, vindictive entrepreneurs rarely succeed.

In our case, we were far from cranky. We were thrilled. Our enthusiasm for becoming rocket scientists was infectious and people came out of the woodwork to be a part of our mission to create a space company from scratch.

There is something strangely compelling about entrepreneurs with a creative fire in their bellies, all-in to do the impossible, which draws those around to help. With Starsys, it was a growing recognition from the folks at NASA that we were going to become a space company, come hell or high water. We were unshakeable in believing that our product deserved to be on spacecraft—if we could just get to the right people about the right problem at the right time.

> **Tibbitts Tip**
> **Practice Optimistic Tenacity**
>
> When a customer or investor or potential soul mate says "no" to something you are confident they would value, a powerful self-response is "I must not have conveyed the value proposition ... as this is clearly something of value for them ... how do I express that more effectively next time?" vs. "I must have been mistaken that what I am offering is of value."

FROM THE GARAGE TO MARS

As our tribe of supporters continued to grow, people began to cheer us on, help us out, and make introductions. They would patiently walk us through our misconceptions and share the secret decoder ring rules to being a space company. A decade later, there were dozens of folks who proudly said, "I was there at the start. I helped these guys get it going." We were pleased to agree, as truly we would not have made it if it hadn't been for those who were rooting us on and forgiving our naïveté.

It was hard. Really hard. The space industry has a ridiculously high barrier to entry. Its single rule for being welcomed into the club is simple: *It is impossible to get your hardware on a spacecraft until your hardware has been on a spacecraft.* In short, you have to have flight history to get flight history. It is a ridiculous premise. As statisticians say, "One mouse is no mouse." Similarly, one successful operation in space is statistically insignificant related to success on your second time in space. But in practical terms, engineers have been fired for being the first to try some device in space and having it fail. No one gets fired for being second.

We asked anyone we could get on the phone what materials were OK in space, which weren't, and what the space hardware rules were. We started building actuators out of the right materials, taking pictures, creating specification sheets, doing everything we could to create the look and feel of an authentic space company.

> **Tibbitts Tip**
> **Create Followership**
>
> Starsys would not have succeeded without creating followership: a tribe of industry experts that became Starsys evangelists.
>
> It was in a part a result of what we did; our innovation, our solving of tough problems. But more importantly it came from our sincerity, humility, desire for human connection, and our commitment to fun and family. We had hundreds in the aerospace industry rooting for us to succeed, and that made all the difference.

BONA FIDE

Then there was the matter of what to call them. Sadly, we resorted to acronyms, thinking that is what space companies did. High-Output Paraffin Actuators didn't sound too bad in the expanded form. But as an acronym, it became a HOP actuator. An embarrassing name, reminiscent enough of rabbits, that we never got past just spelling it out. "Aitch-Oh-Pee" actuators are what they became and have remained so for 30 years.[1]

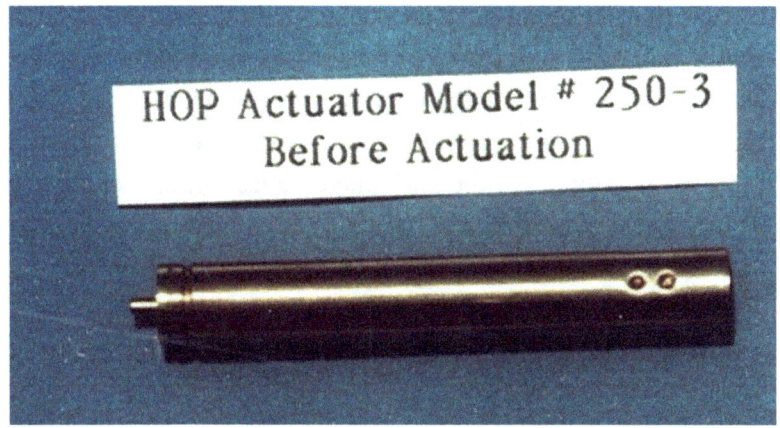

But HOP actuators they were, and we built them out of space materials (Nylon! Epoxy! Titanium!), polished them up to look like jewelry, and tested them. We printed the performance specifications on sheets of paper, took pictures of the actuators, and developed them at Walgreens. Once we pasted the pictures to the specifications with a glue stick, they became product sheets that we sent to every space engineer for whom we could find an address.

---

[1] As Yoda might say, "Product branding, carefully you must choose." The name that sounded so good during that Friday afternoon beer-fueled naming brainstorm needs to sound just as great 15 years later. During the same time, poor branding also was at play in my personal life. We had formed a rock-and-roll garage band to blow off work stress. The chosen name of "Midnight's Crew" was hilarious for a week but quickly became embarrassing for audience and band alike every time we introduced ourselves. The name was quickly changed to "Too Much Fun"— a robust moniker that has stood the test of two and a half decades.

FROM THE GARAGE TO MARS

Thinking back, we were delusional. In our heads, there was never any doubt that we would eventually find someone who would want our HOPs, and so we went merrily on, cavalier to almost a year passing without any real interest, sure that next week would be the one. Not thinking through that it was likely our passion would outlast the cash in the bank.

Then Dick called.

Dick Casper worked for a company called Aeroflex. He was flying[2] an instrument that needed something to push a latch very gently with very high force, and there was nothing that had yet been used on spacecraft that did exactly what he needed. He was stuck. He either had to use an untested technology or he wouldn't be able to fly his instrument. When he called, he had a product sheet of new device on his desk with a picture partially glued to it. I later learned that he was wondering at the time if it was worth the risk, but had decided to call anyways.

We talked, and he asked a few questions about how our devices worked but more importantly, I could sense, he was judging the mettle of me and my company. He shared that we were his only solution, and he was faced with the dilemma that the only way his instrument would work was if he would put this thing of ours into space for the first time. After a pause, he asked the question no one had asked before, and that frankly, we were unprepared to answer:

"How much would it cost for three of your HOP actuators?"

In my excitement, I came up with a number in my mind that would make us bucket loads of money. I was about to blurt it out, but reason told me to shut up, and I decided to wait. It is fortunate that I did. I had no idea how much the number chosen in that moment would impact the company for the next 20 years, as HOPs

---

[2] "Flying" is space speak for "Putting it on the top of a rocket and sending it to space." The space lexicon is rife with understatement. For instance, a launch vehicle blowing up on a launch pad is "an unexpected exothermic anomaly."

ultimately became the bread and butter of our company for a few years. To put this in perspective, if I had quoted a number $1,000 lower, the impact on the company in the coming decade would have been millions of fewer dollars to the bottom line.

"Dick, I will need to do an accurate costing. I'll get back to you. Give me a couple of hours."

I hung up the phone and immediately called Robert Intout, one of the mechanisms engineers at JPL who had taken us under his wing.

Tibbitts Tip
Don't Negotiate With Yourself ...

When pricing, the tendency can be to say, "I don't want to give them sticker shock. Let's give them a lower price that won't offend." As a result, I'm sure many of our customers said, "We would have paid twice that!" The question to ask is not, "Am I asking too much? but rather, "Am I fully conveying and capturing the value I'm providing to the customer?" Tempered by "Is my asking price so high that they will walk without negotiating?" There is always room to sharpen pencils if the customer has sticker shock.

A powerful hack to suss out the customer's price tolerance without offending is to break down the deliverable into a detailed list of the individual elements. Include things that aren't requirements but "desirements." Then describe and cost each individually, rolling them up to the full price. (Think the sticker price of a new car.) If the customer objects, it's easy to remove the bells and whistles to get to a price that works. If the customer doesn't flinch, there is likely additional money on the table.

The Corollary:
Profit is not a Dirty Word

Profit is the fuel that allows you to be financially healthy and have the resource to serve your customers. Do not price on what you think is an agreeable profit percentage. Price on value. You will at times be making 50%. Be proud of that rather than embarrassed, as you've created a stronger company to serve your customers.

"Bob, I've got a customer who wants to buy our actuators. How much should I sell them for?"

"Well ... in approximate numbers, by the time you get through with testing, the explosive squibs you are replacing cost about $5,000 a piece, so that might be a fair price."

I was floored. These were the devices that we had built from hardware store parts for $7.00. With spacecraft materials and all the testing, I couldn't imagine them costing more than $500 to manufacture. I called Dick back. I could not believe that the costs that Bob and I had discussed would ever work, but I was willing to give it a go.

I did my best to control the waver in my voice and faked an arrogant confidence in our pricing. "Dick, we ran the numbers, and including the testing you are talking about, the cost for three actuators will be $16,187." I was sure the additional digits would indicate a detailed costing and minimal profit.

There was a long pause while I waited for his response, sure he was feeling extorted.

"Sounds good. Get me an invoice. I'll send you a check for 50 percent. When can you get started?"

I was stunned. Several years later, I realized I had left $25,000 on the table, but that didn't matter at the time.[3] We were in the space business. I hung up the phone and immediately started outlining what we would do with the $14,500 in profit we were about to make.

I was out-of-my-mind ecstatic. Call everyone you know and tell them ecstatic. Track down that third-grade teacher who said you were never going to turn out to be anything and tell her what just happened ecstatic.

---

[3] The right answer: "Dick, that will be $16,187 for three of our basic actuators, but did you say you needed ½ in of stroke? Well then, the costs to modify it to your specific performance and test requirements will be an additional $25,000."

BONA FIDE

Start-up entrepreneurship is characterized by something rarely discussed, as frankly, nobody wants to hear about the dark side best described as the Valley of Despair. The self-doubt and fears that later in the Journey would wake me at 3 a.m. in the morning with a shot of adrenaline: "I've got ten people working for me and I can't make payroll next week," "Our $2M contract was cancelled, a dozen people have to go," "The bank is calling our loan. We need to find $1M in two weeks."

Entrepreneurs seek each other out for consolation, as most people cannot relate to these dark moments. Over a beer or coffee, they check in with each other: "Are you getting exercise?" "Any good ways you've figured out for getting to sleep again after you wake up panicked at 3 a.m.?" "Finding time for your wife and family?" "Imagine a job you don't have to think about after 5 p.m." (followed by a wistful look and a long pause).

But this is all put up with for the reward: The ecstasy when it _goes_. When this thing that you carved from nothing, then an idea suddenly starts to take shape. The lows are lower than most people can fathom or tolerate. But the highs...oh, the highs! These are the

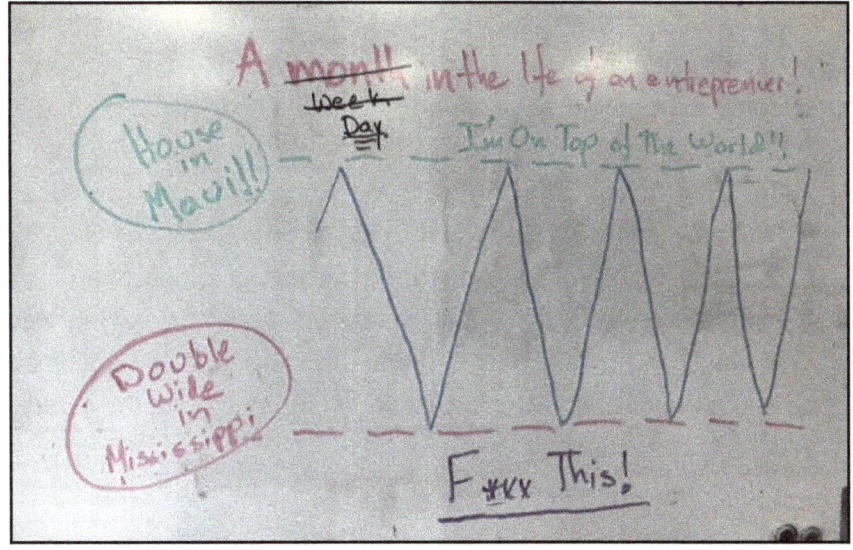

moments you live for—the moments when you open that email or get that call and you realize it is actually *happening*.

This proved true throughout the coming decades of my entrepreneurial ventures. A case in point being this graphic I spontaneously put on the whiteboard during a period of what I came to call SBD, Situational Bipolar Disorder. I have yet to meet a successful boot-strap entrepreneur who doesn't look at that image and say, Yup, nailed it!"

Don't let anyone fool you as to what motivates the true, dyed-in-the-wool entrepreneur. It is not money. Or being in charge. Or being famous. It is those Edison moments; when one sits transfixed for 12 hours straight, watching a light bulb with a carbon filament glowing, knowing you should go to bed but never wanting the moment of "I've created this and it is going to change the world" to end. Most people hear the story of Edison and think "he must have not been able to afford help." Nope. He wouldn't have handed that moment off to anyone for anything. THAT is the juice that keeps us coming back and saying, "Let's do it again." It is an affliction that can shorten lives and damage families, this hunger for the "It Is Alive!" moments.

Funny, but I don't really remember how it felt. Sometimes we truly can't remember emotions—joy, sorrow, pain—but can only remember their impact. The feeling is real and overwhelming in the moment, and then it's gone. Like a picture you take of a jaw-dropping, life-affirming sunset, and somehow when you come back to it, the magic didn't make it to the picture.

What I do remember, though, is walking on Cloud Nine into the Flatirons Athletic Club that evening to work out, having just found out I really was going to be a rocket scientist. I remember walking into the locker room and thinking to myself that there must be hundreds of people in this building, and I knew, absolutely, positively, there was not another person in this building who was as filled with joy, as happy to be alive, and as full of I can't believe this is happening to me ecstasy as I was at that moment. It was just not possible for anyone else to feel this good.

* * * * *

We were supposed to ship our actuators on January 6, 1988. We worked like crazy through the holidays to be ready on time. On January 5, they were complete except for a final overnight test. I needed to leave them in a small oven to confirm that they wouldn't extend spontaneously if the spacecraft sat on the pad in the hot sun.

I put the actuators in the back of the oven with a thermometer in the front, set it to 150 degrees and left for the night, not realizing that the back of the oven was 20 degrees warmer.

I walked into the Valley of Despair the next morning when I opened the oven to three broken actuators. As the paraffin melted and expanded from the over-temperature, it had nothing to push against. It was as if the HOPs had become angrily flatulent, squirting wax in a messy puddle at their feet. We were sure that we were done in the space business. I called Dick with the news. His response was unexpected.

"Very sorry to hear...been there, done that! Not a problem. We can work around that. Can you get them to me by the end of the month?"

We rebuilt the actuators and had them out the door a week later. My profit calculations turned out to be premature. Because of that "whoops," our first contract cost us $32,000 to complete. I had to sell my Triumph TR-6 sports car and max out Jackie's and my credit cards.

FROM THE GARAGE TO MARS

This is a picture dear to me of four of us wearing huge smiles, the actuators in front of us sparkling in the fluorescent light, taken just before we dropped the package off at FedEx, a drawing of the water heater that started it all at the far left of the picture.

I no longer had a sports car. My credit cards were maxed out. I could not have cared less.

Starsys was *bona fide*.

From L: Daryl Maus, the author, Mike Schnettler, Hank Brown.

WAX IN SPACE

WE MAY HAVE BEEN *BONA FIDE*, but that did not make us rocket scientists with full privileges quite yet. It's not broadly shared outside of NASA that when you are selected to be an astronaut, you are not truly considered an astronaut by the rest of the astronaut corps until you've flown in space. Similarly, after we completed our first contract with Dick Casper, we had delivered flight hardware, but our stuff had not been to space yet. Potential new customers were appropriately cautious as they sniffed around our new technology.

What kept the industry interested was the fact that what our HOP actuators could do was unique, solved a Big Problem, and we were a contagiously fun bunch to be around. There was growing interest and awareness of our devices, but customers remained hesitant. It seemed particularly so for the company closest to us, Ball Aerospace. Ball, a thriving space company in Boulder, was not willing to trust a couple of guys in a garage down the street making space products out of wax.

We had time on our hands as we developed the next customers, so Friday afternoons spawned morale hours, with a few of us playing roller-blade hockey or laser tag on alternate Fridays.

Late one Friday in 1989, we were hanging out in shorts and flip-flops, winding up the day covered in the funk of an hour of roller-blading, when we heard a knock on the door. Nobody knocked at Starsys. Those we needed to meet with knew us and just came in. We ignored it, figuring it was someone selling copier toner or wealth management services. A second knock came, and after a pause, a hesitant third. I thought "What the heck. I'll take a look," and opened the door. A thin, tall, gray-haired man in an ill-fitting suit was walking away. I wondered if I should bother.

"Hey, there! Can I help you?"

The man turned, paused, and looked with curiosity at this strange fellow at the door, wearing bright yellow Richard Simmons shorts and flip-flops, sure he was in the wrong place.

He responded hesitantly in a strong, academic German accent, "Yes, maybe. I am visiting Boulder and looking for the company that makes this invention called High-Output Paraffin Actuators. Do you know where this company might be?".

"You found us. Come on in!"[4]

Dr. Klaus Wilhelm turned out to be the Principal Investigator (PI) for an instrument called SUMER that Ball Aerospace was building to fly on the SOHO solar research spacecraft. As the PI, Ball Aerospace was under contract to Dr. Klaus to build the instrument, and as a kick-off for the program, he was visiting from the Max Planck Institute of Solar Research in Munich. He had read of our actuators, coincidentally also made in Boulder, and at the end of his meetings, on a whim, dropped by to see the company and the product.

From his expression (whimsical curiosity) and countenance (nothing else to do, might as well hear these guys out), it was clear that Starsys was not what he expected: two offices, a couple of sweaty guys in shorts, and an odd-looking man who popped out of a back room mid-meeting wearing pajamas, smelling of sleep, and headed to the bathroom with toothbrush and razor.

Daryl and I quickly assembled a demo, putting the devices in hot water and lifting weights, while talking about the product in

---

[4] An echo of this occurred 25 years later in my subsequent venture, Katasi. A phone call, nearly ignored as "another toner salesman," was instead the producer for Katie Couric asking if they could fly to Boulder to do a segment on our distracted driving technology, Groove. Once again, a single connection almost missed changed our future.

general. Within 15 minutes, our enthusiasm for what we were doing, as well as the coolness of the technology began to wear off on him. By the time he had left an hour later, he was committed to using our devices wherever he could on the instrument.

We learned something counterintuitive that day. The farther you are geographically from your customer, the more legitimacy you are bestowed. The same magic that catalyzed Cheap Trick's "Live at Budokan" rebirth. This proved a powerful tenet through the years, as we found that by reaching out internationally to customers, it created an "If we have to go that far to find them, it must be good stuff" cachet.[5]

Klaus went back to Ball and changed our lives. He directed that every device on his instrument that could use wax, should use wax. It quickly became a very different world with the customer now asking us "Where else might we use these things?" Those were heady times, and a potentially dangerous technical arrogance began to develop on our part that perhaps we could do much more on the spacecraft than we had originally thought. But arrogance comes with confidence, and with our customers suggesting we could be more than a wax actuator company, we started to believe we could build the actual latches, hinges, and cover actuators we were being asked to design.

Along with moving up the food chain to spacecraft devices came a substantial increase in the value of what we were delivering. Our typical contract size quickly grew from $15,000 to $150,000, allowing us to bring on additional help and grow the company.

With the instant credibility from Klaus' stamp of approval, we were inundated with requests to provide actuators and various

---

[5] Another echo: Early on with Katasi, we needed to raise $1M and were proposing a company value of $5M ($1M would garner 20% of the company). The Katie Couric segment attracted international attention, including that of Australia. This led to my being asked to come to Sydney and present to investors. The fellow making the introductions explained that I couldn't raise money at a $5M value; it had to be at least $10M, as we wouldn't be coming all the way to Australia for just a $5M company. I flew back to the states with the $1M, while only having to give up 10% of the company.

devices for the spacecraft, eventually leading to more than twenty devices flown, the variety of which warp-drove us into becoming much more than just a wax actuator company.

But it was almost not to be.

At the heart of this work was an ambitious product for the SOHO MAMA instrument, where we were to build an actuator that acted like a brick-sized ballpoint pen on steroids, to both extend and retract when pushed by wax, allowing the main cover of the instrument to gently open and close. It provided more than 200 pounds of force as it toggled between the two positions. The project was immediately and irreverently named Yo MAMA by our technician.

A second tenet surfaced from the Yo MAMA project that would vex us for years as we wrestled to fully comprehend its gravity: "Just because you can convince a customer you are competent to do something you've never done before; doesn't mean you are." Nevertheless, we jumped into the program unencumbered by neither doubt nor uncertainty. Daryl was now our lead designer and his office desk was scattered by the detritus of dozens of disassembled ball-point pens as he became expert in the workings of their latches.

Yo MAMA turned out to be a beast. The "click" of the mega-ballpoint pen was more of a "thunk," causing the cover to fly off its tracks and self-destruct. We tried modifications to the design to soften the click, and each created the same result, reported by phone call from the customer, "The cover has again failed."

We were given a final $10,000 to fix the design. We delivered the test hardware to them on a Monday and were told that if this final version did not work, the customer would be forced to terminate our contract and use a more conventional solution.

We had grown to a dozen employees by this point. We all gathered at a picnic table outside the building the following Friday afternoon to wait for the call. It was clear that although there would be no hard feelings from the customer if Yo MAMA failed once again, it would mean the end of the wild ride. We would devolve to making only wax actuators that our customers were sure we could deliver, rather than mechanisms that required sophisticated, untested designs.

There was a tension in waiting for the call that was strangely exhilarating for all of us, knowing that the call might mean the end of all we had been working for, or it would light a rocket to take it to the next level. I would wager most entrepreneurs are fueled by this corporate Russian Roulette; the future coming down to the turn of a single card, whether a meeting, a phone call, or an email, the juice coming more from the flip of a coin decisions rather than the slam-dunk announcements.

Someone had run out for a bottle of Jose Cuervo, now sitting unopened on the table, and another had rounded up a gross of bottle rockets they had picked up on a recent trip to Wyoming. We were going to have a party one way or another.

I remember the warm sun, the worn brown of the picnic table, and the scent of the surrounding Austrian pines as the call came. Yo MAMA had worked flawlessly. The customer was thrilled, and they wanted to start the order to build the units for the spacecraft.

Tequila shots were slammed, bottles rockets were launched, friendships were furthered, and loyalties solidified. The rosy glow of "We did it!" settled over all of us. To memorialize the event, a tradition was declared in which tequila and bottle rockets would be a part of all future product celebrations.

The tradition morphed in the coming years as political correctness and a visit from the police manifested. In later years, rather than bottle rockets, an Estes rocket built by the product manager and signed by the product team was launched. Rather than tequila, the launch was preceded by a champagne Dixie cup toast to those who helped make it so. Each champagne cork was then labeled with the program name and put on display in the lobby, with the model rockets, signed by the program team, adorning the program manager's bookcases like soccer trophies.

The Rocket Launch became one of several touchstones of the Starsys' unique corporate culture that was quickly developing, reminding people of our roots, our unique ways of saying "well done" and the importance of patting ourselves on the back now and then. This, and other traditions, began to codify an edginess in the company that became an important part of our success, attracting the very best talent to the company as well as the best customers.

Stories spun out from our Rocket Launches over the next two decades that became company legend: The rocket that was expected to glide gently to the ground, but instead augured in under full power into the only employee vehicle in the lot that was worth enough to warrant repair.[6] Or the Lockheed Martin customer that declared, Crocodile Dundee style, "That's not a model rocket. *THIS* is a model rocket" and presented us with a 7-foot behemoth that he had personally built to celebrate his program completion. A rocket that we found out later was far too large to be launched from a parking lot shared with the Boulder Country Day elementary school.[7]

The Rocket Launch became a reminder of the specialness of our company. The best traditions should be retired before they get stale, replaced by new traditions, fresh and relevant. But the Rocket Launch was beloved by the company and lasted a long time. We launched hundreds of model rockets over 25 years. When I finally

---

[6] The Boulder Body Shop dent repair 275 line item puzzled our accountants.

[7] Someone (perhaps the author?) naively suggested using a smaller motor for the launch, unaware that this would create the unintended result of providing less than adequate altitude for the parachute to deploy. The rocket reached about 400 feet before arcing over and descending lawn-dart style towards the parking lot at the back of the building. We were sure we would find a car punctured through the roof on the far side attended by a hysterical parent, but to our good fortune, the parachute had opened just after disappearing behind the building.

First four on the left: Starsys Research Team. Five on the right, Lockheed Martin Team, Kyle Peter, Program Manager, 4th from left.

decided it was time to retire the tradition with one final rocket launch, it was 2006 and something I would personally oversee.

I had not been involved with building the rockets for more than 15 years, when I went into Hobby Lobby and purchased that final Estes rocket. I put much care into building the rocket in our garage, gluing the fins just right, carefully spray-painting it a robin's-egg blue on the nose cone blending to a canary yellow at the fins. The next day, I walked through the company with a silver Sharpie, asking for signatures and crowding more than a hundred names on to the rocket's body.

On the last day before we moved to a new facility with twice the floor space, we stood in the parking lot and toasted the thousands of mechanisms we had built in our building over the past ten years. After a chorus of *5, 4, 3, 2, 1*, I pressed "fire" and the rocket streaked into the sky with a hundred employees shrieking with the delight of children. The rocket now sits in my office with a significance few fully understand.

The final Estes rocket with more than 100 names of Starsys employees.

CHOOSING THE DNA

When you start a company from scratch, you are hunkered down, putting all you can into the singular focus of carving out an entity that you hope someday will become self-sustaining and thrive. At the same time, and usually without conscious intent, you are establishing and locking down the norms and principles the entity will live by. All of the Right's and Wrong's as to how the company, and those within, shall act. Codifying the behaviors that will be celebrated and acknowledged, as well as those that will be disparaged and discouraged. As with Dr. Frankenstein, there can be so much effort put to bring the creature to life, that the personality of what is being created is given short shrift.

Whether given conscious thought or not, corporate culture forms primarily as a reflection of the founder's personal values, projected into an organizational code of ethics—the company DNA that will guide the hundreds of decisions made by the firm every day. As the organization grows over time, these values become immutable stanchions driven deep into bedrock.

A hologram is a three-dimensional image, characterized by the fact that any piece of a hologram, no matter how small, contains the complete image. Organizational DNA is like this. Spend ten minutes talking to someone in any company, about the company, and you can pick up on the values by which it operates. Or, simply walk through a company, look into the cubes, see the cartoons on the wall, listen to the conversations, and again, the DNA is manifest.

FROM THE GARAGE TO MARS

Not all personal values of the founders should be branded into the company. We quickly learned to intentionally discriminate founder values that should be nourished in the company from those that needed to be quarantined.

After five years, Daryl sold his portion of the company back to the company and went off to continue his personal entrepreneurial journey. It was now up to my values to inform the culture of the company.

In my case, I had strongly held values around *Excellence, Innovation, and Integrity* that were appropriate to let run rampant. So far so good. But at the same time, some principles I held were NOT the stuff a space company should be made of. My entrepreneurial valuing of risk-taking over caution, good enough over perfection, and spontaneity over planning, could have created a dangerous mojo within a company making a product that must never fail.

> **Tibbitts Tip**
> **Company Values—**
> **Carefully Choose You Must**
>
> Some of the entrepreneur's personal values are rocket fuel for the company mission; others are sand in the gears. Objectively select which of your personal values to let loose into the company, and which to temper.

After a few early embarrassments (the overheated actuator debacle of our first actuator build, for example), we wisely brought other leaders into the company who lived, ate and breathed the values that I did not. I recused myself in these areas and empowered them to drive their key values deep into our organizational DNA.

We chose early on to not only be intentional in selecting our DNA, but to consider Starsys a grand experiment. We wanted to establish an organizational culture that would create great results by operating at the edge of convention by declaring a commitment to values heretical for an aerospace company. Values that celebrated *Family, Fun, Quality of Life,* and *Work-Life Balance*. The reasoning

CHOOSING THE DNA

was not a profound parsing of organizational development theory. We simply were creating the company we had always wanted to work for. We did not recognize at the time that these four values would become the key to our extraordinary success.

Pretty quickly, we found that a critical part of replicating values in a company was to hire those who pre-resonate with that value set. This created quite the challenge in hiring, as we found that not only could we not afford to hire aerospace technicians, but they were often also a poor fit for our culture of flip-flop sandals, fun, and family.

Instead of hiring space folks and trying to PlayDoh them through a values Fun Factory into a culture they were uncomfortable with, we discovered a hack. Find the diamonds in the rough pre-wired with our values and teach them to be space folks. Hire for cultural fit and teach skills vs. hire for skills and teach values.

We learned that if we wrote ads that oozed our values, these would puzzle many and be a siren call for just the right people:

> *"Are you tired of flipping burgers? Have you always wanted to build stuff that goes into space? Are you a wizard with your hands and can fix just pretty much anything? Come by and meet Starsys at 5757 Central Avenue".*

When we put out an ad like this, we would quickly dismiss mailed resume responses. We were looking for more. We were looking for people who came to our front door, a glazed look in their eyes, mumbling, "How did you know... that ad was written for me... I belong here!" When we saw that response, the interview process was brief. To their amazement and delight, we usually said, "You're hired," pulled them into the company, and taught them how to make space stuff.

FROM THE GARAGE TO MARS

Of the heretical values we held dear, at the top of the list was the "F" word—Fun. At Starsys, we took our Fun very seriously. Fun as in spontaneous work breaks where we made popcorn and watched old-school Chuck Jones cartoons. Fun as in making ice cream from liquid nitrogen every Friday afternoon. Fun as in every new team member having to share "Two Truths and a Lie" or "My Most Embarrassing Moment" on their first day of work.

But the most Fun EVER was the Vomit Comet.

When I was ten years old, I had read of a reconditioned commercial airliner that NASA used for astronaut training. The seats had been removed, the walls had been covered with foam, and NASA flew it out over the Gulf of Mexico in grand parabolic arcs to provide those within the experience of zero gravity.

The plane was the ultimate roller-coaster ride, and in third grade, I declared that someday, somehow, I would find a way to fly on that plane. The dream lay dormant for two and a half decades.

Tom Zawistowski was a space engineer from the NASA Goddard Spaceflight Center. He called out of the blue one day shortly after we had delivered our first few actuators asking for a favor; he was testing an antennae device on the NASA Reduced Gravity program plane, the same plane I had read about long ago. He was short on funding and needed to borrow an actuator. He proposed a trade—for the use of our actuators, we could then say our actuators had been testing in zero gravity. The request was like a seven-year rain triggering locusts to re-emerge. If I would have been in a cartoon, a light bulb would have appeared above my head as the long dormant wish-ember flared.

I suggested a different deal. If they had the room, how about I come along as the actuator-wrangler? He paused and through the phone, I could feel him cocking his head and pondering. "Yeah, that could work. You'll need to get high-altitude training for the plane, but I'll put you down as a 'contractor test technician.'"

CHOOSING THE DNA

Six months later, I was in the belly of NASA's Reduced Gravity Program KC-135,[8] 20,000 feet over the Gulf of Mexico. The plane was known by all, without a hint of irony, as the Vomit Comet. It was a nod to the physiological response to being in a plane that cycled through half a minute of zero G immediately followed by two minutes of being plastered to the floor of the plane by two times the force of gravity. Forty times in a row. The name arose from the fact that two out of three passengers became violently ill from motion sickness.

There was no "Excuse me, Mr. NASA ... I'm feeling a bit queasy...could we go back to Houston?" Once the roller-coaster started, there was no getting off for two hours. If you were going to be sick, you were to grab a barf bag from the breast pocket of your flight suit, deal with it, and get back to work. During the pre-flight briefing, the flight directors made it clear that throwing up was just part of the adventure, but to miss your barf bag in zero G was poor form, as it would create a breakfast jellyfish floating through the fuselage that would get you permanently banned from the program.

I was standing mid-way down the fuselage with ten other engineers, each of us with our barf bags peeking out of our flight suits like handkerchief squares ready to be whipped out like a parachute ripcord at the first hint of a breakfast expulsion event. The NASA test engineers said, "Here we go!" and the plane dropped into a 45-degree dive toward the ocean, plunging 10,000 feet in 60 seconds, pulling out of the dive. And gravity went away.

Really. It literally just went away. You know how it feels when you are on a roller-coaster and it goes over that first hump and you feel your stomach drop out from under you and it creates that put your hands in your air, scream like a little girl, isn't this fun feeling? This was that at first...the feeling of going over the hump and that "whoa!" thrill as the gravity lessened. Except that instead of the

---

[8] A Boeing 707 with the seats removed.

reassuring return of gravity, it—just—kept—on—going—away! It was like someone had been given the gravity dial and they just kept turning it from 100% to 50% to 25% to 0% over a period of about five seconds. Despite my commitment to not acting like an idiot, there was no controlling the "Whoooaooowhoooo!" scream of joy that came from me as gravity disappeared and I floated up off the floor. I thought zero gravity would be really, really cool. It was sooooo much cooler than I had imagined.

It was nuts at first trying to figure out what to do to control the motion of my body. Outside of conscious control, my amygdala freaked, commanding my limbs to flail wildly, trying in vain to maneuver my way back to something solid that I could grab. But physics doesn't work that way in zero G. Your body merrily floats along in whatever direction it was going when gravity went away, ignoring akimbo arms and madly bicycle-pedaling legs, until you eventually encounter the other side of the plane. It matters little what you do between A and B other than how ridiculous you look to others in the plane. I quickly learned to relax and enjoy the flight.

Zero G is strangely familiar and comfortable. Something that feels like it's the way we were always meant to be. Maybe it's a subconscious déjà-vu of floating in an amniotic sack, or a reminder of floating quietly like a jellyfish in the neighborhood pool, but for me, the familiarity came from somewhere else that I did not discover until a later flight.

I was in the back of the plane with little to do as our testing had ended. I asked the flight engineer if I could go to the cockpit to see what it looked like from the front when a commercial airliner points alternately 45 degrees at the ocean below, and then 45 degrees at the sky.

"Sure!" he said.

"How best to get up there?"

"Fly!"

CHOOSING THE DNA

"Huh?"

"Yeah … do a Superman!" (You can't make this stuff up. That *is* what they call it).

As gravity disappeared on the next parabola, I pushed off from the top of a seat in the back of the plane, and floated along the ceiling toward the front of the plane. Imagine sitting in a commercial airliner and seeing the flight attendant floating from the back of the plane to the front, along the overhead bins, and you get a sense of the surreal-ness of it all.

The plane was filled with a half-dozen researchers below me, intent on doing their experiments in the 25 seconds they had of zero G, while I floated above them un-noticed. Half-way to the front, it struck me that I had done this before. In dreams I had as a child, where I could effortlessly fly over the countryside, seeing everything below, unencumbered by gravity and the limitations of being rooted to the ground. A pure joy of being completely unencumbered. This feeling was that. Experiencing the joy of flight as if a bird. It was truly wonderful, and added additional depth to my experience that "Dreams really can come true."

Near the end of this flight, I talked the NASA photographer into taking a group photo. It was the last parabola of my final time in the plane. We organized ourselves into a small flock of engineers, some of us upside down, floating in the cabin, smiling at the camera. The cameraman picked up the camera, pointed it at us...and the lens floated away! He grabbed it out of midair, as my last few seconds of zero G, ever, ticked away. He spun it into the fitting, swung it wildly in our direction, and just as the gravity began to return, took the shot. A second later, gravity came back full force, and we collapsed in a heap on the floor like four sacks of potatoes, howling with laughter. The picture became the centerpiece of an article in *Scientific American*, and is now in a frame above my desk, a reminder of an experience I'll never forget.

A Vomit Comet Starsys Research team. From upper right clockwise: the author, Mark Richardson, Jason Priebe, and Scott Christiansen. This photo was taken three seconds before gravity returned and the four researchers landed in a heap on the floor of the plane.

It turned out I was one of the lucky ones who did not get nauseous. Flying on the Vomit Comet was just pure Fun. True to our values, we had to find a way to share this experience with the company.

Once I had met the people in charge of the Reduced Gravity program, I started to explore how we might be able to arrange our own Vomit Comet flights. We actually had a valid reason for doing so; many of our devices we knew would act differently in zero G and needed to be validated they would work properly. This led to

CHOOSING THE DNA

the Vomit Comet Pitch. For fifteen years, near the end of a contract negotiation with a customer for some multi-million-dollar space mechanism, the dialog would go something like this:

> **Scott:** "You know, we really should fully test this in zero G. The best way to do that is on the Vomit Comet, of course."
>
> **Customer:** "Wow, I didn't know that was an option. How much would that cost?"
>
> **Scott:** "Well, with set-up for testing, training of the techs, two days of flights, it would cost $XXXXX."[9]
>
> **Customer:** "I know it's important, but I'm not sure that will fit into our test budget."
>
> **Scott:** "And...of course...that cost includes the Customer Observer, who will need to come with us for the flights, which would likely need to be you."
>
> **Customer:** "You know, that IS an important test protocol to include in the program. Please include that in your final costs."[10]

Once we had the commitment to test, the way it worked was that Starsys could take a max of five technicians for each experiment. Strangely, the Starsys team always seemed to need all five, one of whom was the Customer Observer. Two were needed for the test itself, leaving two additional slots.

These were filled by raffle. Anyone who wanted to go on the flight—whether it was a vice president or the person answering phones at the front desk—could put their name in the hat. Names

---

[9] Actual amount redacted to avoid "Space Hammers cost $500!!!" effect.

[10] A key part of any sales value proposition is knowing your customers wishes better than they know them.

FROM THE GARAGE TO MARS

were then drawn to fill the two empty slots. The reason? We wanted to give as many people as we possibly could the experience of what it was like to be an astronaut. A typical company would have limited this to a perk for higher-ups, but we wanted to make the point that we were all in this together as a team. Whether you turned screws on hardware, or were the head of engineering, we were as equally committed to you having a working-at-Starsys experience that **Rocked**.

And then we came up with a killer way for us all to share vicariously in the experience of the team's zero gravity experience...the Vomit Pool! (Maybe we should have come up with a more politically correct name?...Nah!). When the team went to Houston, we would put a poster up on the common-area wall, with a matrix of each of the team members and the days they were flying. Every employee who wanted to participate could put $5 in a bucket and vote for who they thought would throw up on which days:

> **Tibbitts Tip**
> **Upset the Hierarchy**
>
> There are plenty of perks for those high in the hierarchy, far fewer for the body of the company. Lessen the resentment of hierarchy by going out of your way to find opportunities for egalitarian rewards, participation, and leadership. Flatten the organization, or if you can, turn it upside down by providing rewards, acknowledgment, and opportunities to lead to those lower in the hierarchy pyramid.

"I think Rebecca will throw up on Monday and Tuesday. I think Jason will throw up on Tuesday. I don't think Matt will throw up at all".

It was our version of betting on the brackets during March Madness. Every morning, the team would report in, "Jason threw up today. Rebecca managed to keep her cookies down." At the end, the person who best predicted the outcome ended up with more

CHOOSING THE DNA

than $100. It became a part of our culture that was most bragged about internally and envied externally.

But there was method to the madness. The key to promoting values in an organization is NOT what you say or what you put on the wall on a poster. It is what you do at the edges that shows you are not just turning the HR values crank. Yeah, many small companies do things to reflect their alignment with fun: put a foosball table in the break room, a volleyball net in the backyard, or the easy cheat, pay to have the summer party at an amusement park.

That said, when an organization is truly all in, it invests leadership bandwidth and focus on how to make this company outrageously Fun, leveraging its Innovation value to serve its Fun value. That's when the value becomes bedrock. It's not just lip service to attract employees. It's in the DNA.

This was also a powerful testament and forcing function for other Values important to us. The power in self-deprecation, allowing others to have some harmless fun as a result of your human frailties, brings a team together from the "We all put our pants on one leg at a time" effect. I crossed fingers that our senior leadership would be the ones on the plane to get sick to emphasize that we are all really the same when it comes down to it.

Our allowing anyone to put their names in the hat put our money where our mouth was when we said we did honor everyone on the team equally for committing their work lives to Starsys. We paid for the two extra people to be on the plane to show that we were serious about every one of our Values. That what we said was important we truly made important. In an odd way, making it clear we were dead-serious about Fun made it easier to land hard on the Values of Integrity, Excellence, and Innovation when we needed to.

FROM THE GARAGE TO MARS

The cost of our Vomit Comet program was expensive. Tens of thousands of dollars were spent through the years, as dozens of Starsys employees got their chance to experience for a couple of hours what it is like to be an astronaut. But as the Master Card ad says, the value of 150 employees understanding that our company was no-kidding-committed to investing in the quality of the lives of their employees? Priceless.

MORE THAN WAX

THERE IS ONLY SO MUCH WAX you can put on a spacecraft.

Morris Maslow was a psychologist best known for his "Hierarchy of Needs" theory.[11] He is also responsible for a catch phrase I heard first from a contracts manager at Ball Aerospace: "When all you have is a hammer, everything looks like a nail." When you are very good at one thing, you view the world through the lens of that one thing that is likely the solution to everything you do.

Our hammer was wax. We saw the world of spacecraft mechanisms through a narrative that went something like:

1) Wax is a good way to move things on a spacecraft.

2) Spacecraft have things on them that need to be moved, therefore

3) Wax is the best way to move anything that needs moving on a spacecraft.

> **Tibbitts Tip**
> **Be an Agent for Personal Greatness**
>
> Without conscious intent, Starsys had become a company that motivated its employees from high purpose—self-actualization—the top of Maslow's pyramid. In retrospect, if there was one simple dictum for Starsys' success, it was this: Hire good people with potential and then work like the dickens to help them become what they always wanted to be.

---

[11] Maslow's Heirachy of Needs theory posits that humans are first motivated by basic survival needs, such as food and shelter. Once these are met, we are motivated next by needs for safety, self-esteem, and love; and when these are met, for self-actualization—to realize our full potential and purpose.

The attentive reader should notice a glaring leap in logic between points 2 and 3. However, the paraffin technology was clever and innovative, and it begged to be applied to more interesting uses than just latches. Dangerously, our customers were also adopting this thinking as they heard of our technology and company. Being new to the space business, we figured if our customers thought it were so, that maybe we *could* use wax whenever something needed to be moved. This came to a head in the mid-90s when we began working with Lockheed Martin on a big, new, high-profile spacecraft.

Beginning in 1991, and continuing through 2003, NASA launched four Great Observatories to allow us to see deeper into the universe than ever before. Each of the four was built to capture images in one of four unique wavelengths: visible light, X-rays, Gamma rays, and Infrared."

The Hubble Space Telescope was the first of these. It was a visible light telescope that would not be hindered by diffraction of the atmosphere or stated more simply, hindered in the same way that air makes stars twinkle. Twinkling is great for romance and children's songs, but it wreaks havoc when viewing a galaxy at the other side of the universe.

The Advanced X-ray Astrophysics Facility, or AXAF, was a space-based telescope being built as another one of NASA's Great Observatories. Later named the Chandra X-ray Observatory, it was placed above the Earth's atmosphere as an orbiting satellite to see things that couldn't be seen on the ground.

Our atmosphere is even harder on X-ray astronomy, as X-rays are completely absorbed by the atmosphere[12], rendering an earth-based X-ray telescope an oxymoron. AXAF was to be the second Great Observatory, the follow-on act to Hubble. A telescope that used X-rays to view black holes and pulsars, the most violent

---

[12] Good thing. Life could never have developed on Earth if the atmosphere did not protect us from the lethal bombardment of high energy X-ray and gamma radiation thrown off by our sun during solar storms. Life as we know it would have been toast.

and energetic objects in the universe. A black hole has too much gravity to let visible light escape, but X-rays have more gumption, being able to escape the gravity wells of these super-dense objects. AXAF would give us a way to see things we otherwise would not. We would finally be able to peer into the workings of a black hole.

The instruments and telescope within the AXAF spacecraft required covers to protect the sensitive optics and other mechanisms during the violence of a launch, and then needed to open once in orbit. In a sense, these were simply lens caps to be firmly affixed at launch and then removed in orbit. However, the lens caps and the devices that moved them were numerous and complex with air-tight seals of various materials, ranging from small covers the size of a saucer to those that were more than five feet in diameter.

And they absolutely had to work.

If even one of these lens caps failed to open, the $1.7 billion dollar AXAF telescope would be blind, unable to see, and could ultimately be rendered useless. On the heels of the Hubble optics problem, a similar catastrophic goof-up on AXAF could threaten NASA itself.

At the time, one of our company's goals was to be a large part of the AXAF program, and as it turned out our timing was perfect. Our devices were working in space and getting much notice. With AXAF being one of the next big NASA programs, we were looking for more ways to be involved. At the same time, the AXAF team was searching for innovative ways to solve problems, such as how to open and close covers with extreme reliability. We were ultimately selected as the technology of choice for the cover system for the AXAF CCD Imaging Spectrometer or ACIS, which, essentially, was a camera at the end of the telescope that would be able to take multi-color X-ray images of objects at the very edge of our universe.

The cover motor was to be built by Starsys, and it would rely on two of our paraffin actuators, one to open the cover and the other to close it. The resulting ACIS device was a brick-sized metal box with a cylinder of wax extending from either end. When one of the wax cylinders was powered, the wax inside slowly melted, building up thousands of pounds of pressure that then turned a shaft extending from the middle of the box with an almost irresistible force that opened the cover. After opening, the opposite wax cylinder could be powered to turn the shaft in the opposite direction, closing the cover. It was far from a fast motor, taking more than five minutes to warm the wax and operate the device, but for something that was going to be used only once in space, that was fast enough.

The program to invent and build the ACIS cover took us more than a year as we had to design the device from scratch. The resulting mechanism was clever; it used high-strength metal bands wrapped around a shaft to turn the high-force, back-and-forth motion from the HOP actuators into a high-force rotation that would open the cover. To fit all this into a small box meant that everything had to be made just so. Widths, lengths, and diameters of the various parts needed to be cut and polished to exact specifications, as an error of only one thousandth of an inch in a single dimension could be catastrophic.[13]

The program went relatively well. Starsys delivered the hardware to Lockheed Martin, supporting their schedule of putting the instrument together and on the spacecraft, and then Lockheed shipped the spacecraft to California for final thermal vacuum testing before launch.

Thermal vacuum testing or T-Vac subjects the spacecraft to extremes of temperature, as well as the vacuum of space, as a final

---

[13] Space motors are tricky stuff. A shaft turning in a bearing could appear to work perfectly. However, when cooled to 40 degrees below zero in space, the shaft shrinks and the bearing the shaft goes through shrinks as well, but the bearings shrink just slightly more than the shafts do. If we were to make the hole in the bearing just a wee bit too small (by a fraction of a width of a human hair), when all gets cold, the bearing will grab the shaft like Tonga Fifita's feared Death Grip, preventing any rotation whatsoever. Oops! A $1.7B spacecraft is now useless.

test of spacecraft systems. The thinking is that if it works in the chamber at these extremes, it is certain to work in space. Performed in huge chambers large enough to hold a school bus, the spacecraft is placed in the T-Vac chamber. The chamber door is then closed, and all air is pumped from the chamber over a period of days or weeks. During this time, the temperature is set so it alternates between -40 and 140 degrees while each system is checked.

By the time T-Vac is performed on a spacecraft, everything is expected to be working flawlessly with the final check being one last way to make sure all systems are a go. The schedule from T-Vac to spacecraft launch is aggressive with little time allocated for correcting unexpected problems. T-Vac testing something this size is crazy-expensive, racking up hundreds of thousands of dollars a day in facility costs.

For cost reasons alone, it is *NOT* a good thing to be perceived as the root cause of a T-Vac failure.

* * * * *

But that is exactly what had happened. Deep into the test sequence, various NASA representatives and customer program managers had turned on the power to our silver box, and waited for five minutes, then 10 minutes, then 15 and ... nothing happened. It appeared that the Starsys device had catastrophically failed in the attempt, and a late-night phone call had been made to the CEO of the company they believed responsible, taking us back to introduction to this story and the impending headline:

NASA trusts small Colorado spacecraft company to help build the next great space observatory. They screwed it up and the $1.7B mission is now a failure

And so there I was, earlier in the day, pulling into a parking lot filled with cars from NASA, MIT, Lockheed, and TRW, surrounded by scientists and engineers anxiously looking over the shoulder of Scott Christiansen, our head of technology. I could feel droplets of sweat trickling down my back, caused as much I suspect from worry as from the heat of the bright lights and bodies squeezed tightly together in our small cleanroom.

All leaned in as Scott removed the final screw, and I steeled myself to meet one of two futures, each one dependent on how much care the technicians and engineers had put into designing and building this silver box in front of us.

The two paths forward for Starsys were laid out clearly from that moment forward: Open the cover to a mess of broken parts, and word would get around quickly in the industry—Starsys was good people but in over their heads so stay away. Or open the cover and discover that our hardware was not the culprit, and there would be highly visible vindication and validation.

We had arrived at a crossroads with the two possible paths presenting very different futures. Each path was dependent on the outcome of this space company's experiment with a device that had been buttoned up months ago by technicians and engineers, who had never imagined their workmanship would be on display to a dozen skeptical customer representatives.

We were at the turn-of-the-card moment that entrepreneurs curse, but secretly ache for. Roll the dice. Feast or famine. Life or death.

In that moment, our emphasis on fun and family, such as playing with yo-yos and superman-ing in the Vomit Comet, felt childish. It was dead-serious business building rocket hardware, and our approach suddenly seemed foolishly cavalier.

It was clear that all in the room expected to find the smoking gun in the next few seconds. A metal band broken, a shaft mangled, or, possibly, a puddle of congealed wax, confirming the tangible, building sentiment in the room that it had been folly to trust a critical piece of spacecraft hardware to a company that built motors made of wax.

With bodies jockeying for position and leaning in, each customer rocket scientist tried to squeeze forward to secure a front-row view of the Great Reveal. The final screw was gently removed by Scott in the grip of his tweezers, and his gloved hand teased the cover off the top with a quiet *smuck*, tilting it back just enough so that the bright lights spilled inside.

I had not seen the inside of this mechanism before. With everything that had been going on at the company at the time, this was my first look inside the device we had created to open an instrument's sensors to the stars.

And...it was beautiful! Drop-dead gorgeous. It was like looking into the back of a Swiss watch. The light bounced off the polished metal parts of silver, gold, and gray so cleanly polished they sparkled. It was like opening the hood of a new Maserati and basking in the craftsmanship. There were no puddles of wax, no metal shavings or bent shafts. There was not a spot of grease or any sign of wear. It was simply a brilliant piece of work that looked as if it had hardly even exerted itself through the dozens of operations it had already been subjected to.

The throng of rocket scientists leaned in closer to peer at this piece of space machinery that all would have been proud to have created. The mechanism's interior displayed a level of craftsmanship that the group was unaccustomed to seeing. The care in creation was inconsistent with a device that just seconds ago they thought had failed to do its job.

The release of tension in the room was instant and palpable. Bodies leaned back with arms crossed in the reflective side discussions that ensued. I imagined someone pulling out a hound's-tooth hat and clay pipe while raising the begging question: "Clearly, Watson, if it is not possible that *this* mechanism is the culprit, what else could be suspect?" Within a minute, all accusatorial notions were holstered, and the group began to talk about other possible causes of the problem.

The initial conclusions of our visitors proved spot-on. As we tested the device further over the days that followed, it performed beyond expectation, and we were able to quickly prove that there must be another root cause.

And one was eventually found. It turned out that the O-ring seal on the instrument cover had over time bonded the lens cap to the instrument so that far more than the specified force was needed to open the cover.[14] Our motor had performed as promised, trying for 15 minutes in vain to open the cover before going into fail-safe mode to prevent further damage to the instrument when the cover did not open.

The O-ring problem was eventually resolved by a simple workaround. Since disassembling the spacecraft and replacing the seal cover would have created months of launch delay and cost tens of millions of dollars, the solution was clear: Wait until the spacecraft was in direct sunlight and toasty warm in orbit to make sure the O-rings were soft and compliant before opening the cover.

With Starsys vindicated and our place in the AXAF program once again secure, we were given a precious reward the following year: Five of us from our company, including Scott Christiansen and I, were invited by NASA to the launch in Florida in July 1999 and welcomed to the VIP section at the Banana River viewing site.

In the year prior to the flight, the spacecraft had been rechristened *Chandra*, the nickname for the Nobel Prize-winning astrophysicist, Subrahmanyan Chandrasekhar, who was responsible for

---

[14] Unknown at the time to the designers of the cover, O-rings, squeezed between two plates and subjected to vacuum and cold temperatures, flow like super glue into micro-fissures in the metal, bonding them together with forces that no motor could have overcome. It was not that our actuator had not provided adequate force; rather, it was that the force needed to open an O-ring-sealed cover was not achievable by any motor. This is an important take-away for readers with space industry ambitions: The next time you are designing covers for spacecraft, do not use O-ring seals against bare metal.

much of the astrophysics that led to a greater understanding of black holes and massive stars. *Chandra* was a fitting choice as it is the Sanskrit word for the moon.

The trip to the Banana River site was our version of spring break. We headed to Orlando, rented a convertible, and got rooms at the Cocoa Beach Hilton. A shuttle launch during daylight is amazing. The night launch planned for Chandra was said to be spectacular. We were stoked. However, the first two trips to the launch site had ended in the disappointment of a launch delay and a late-night trip on a quiet bus back to the hotel.

But having to spend a handful of days in Cocoa Beach due to launch delays was far from serving hard time. The five of us spent the days between launch attempts drinking Coronas while relaxing on beach chairs with an occasional break to cool off and try body surfing.

After the two delays that were spread across four days, it was time for the third try and we headed to the NASA bus parking area just before midnight, top down, five excited engineers up late and ready for the big event. We drove to the parking lot, hooting and hollering, pulling up next to a large group of well-behaved, comparatively subdued senior space executives. Their decorum was in sharp contrast to the Starsys team—a difference that was pretty much noted by everyone in attendance. But rather than being judged, it seemed that our overt sense of fun was not perceived negatively but with envy and that we were being regarded with a quiet admiration, something akin to "Those are the Starsys guys who make that great spacecraft hardware!" After getting out of our convertible, we hopped on one of the tour buses with the enthusiasm of five-year-olds riding to kindergarten for the first time.

So ... after all the waiting, it finally, actually, happened. And it was far more spectacular than I could have imagined. The following

FROM THE GARAGE TO MARS

are notes I wrote on the flight back the next morning to help share the experience with the rest of the company.

10:00 a.m., July 23rd, 1999, United Airlines Flight 1079

It is Friday morning and Chandra launched last night. The attempt Monday night was scrubbed at T-6.5 seconds because of a small hydrogen "burp" that happened at exactly the wrong time. Forty-eight hours later we waited in the heat and humidity for thunderstorms that never settled down. The third try was the charm.

Last night, with the help of a clear sky and a flawless countdown, it went! We arrived at the Banana River site two hours before launch, just about as close as you can get to the launch pad. As the launch clock worked its way down, we wandered around the site passing the time. That abruptly changed when the clock passed T-5 minutes, the last weather hold point, and it all became real. Our focus shifted to the shuttle, loaded with the energy of a small nuclear weapon, and brightly lit with five people perched on top. Not another soul within three miles of them. There is a strange and marvelous feeling that comes to a person seeing the Shuttle on the pad so far away ... realizing how dangerous spaceflight is and how five representatives of humanity, not drastically different than you or me, are risking their lives to ride the flaming beast into the heavens. It brings to life the words Mission Control said to John Glenn just prior to ignition of Friendship 7: "Godspeed, John Glenn."

You cannot help but feel in the moment a visceral, cathartic connection to those five crew members, representing mankind's willingness to risk it all to explore further.

When I've seen a launch on TV, the risk is abstract and my casual reaction—"Sure, it'll make it!"—is no doubt at least partly a byproduct of my more detached connection to the event. Watching it in person, though, the risk is very real; a 1% chance of catastrophic failure that will blow everything to bits. "Please let that 1% not be this one" is the heartfelt petition that is offered as if a prayer.

At T-2 minutes, the lights go out and 2,000 people fall silent. Not much happens during these last two minutes, just a few short terse comments over the PA between the launch director and the shuttle commander. At T-6 seconds, the main engines fire and you realize that this thing is really happening. Then the solid motors

MORE THAN WAX

light up and more pure power than you could ever imagine is happening before your eyes and ears, and you are suddenly witnessing a brilliant, blinding display of light across the landscape as if the sun itself is rising off the pad. The yells and whoops seemed to come out of my throat from nowhere; in that moment, I felt a small part of me was going up there with them. Powerful stuff.

Chandra is the first Great Observatory to widely use Starsys technologies. As with other programs of this scope, the hardware was delivered several years ago and since that time our attention had shifted to other tasks at hand. But Chandra is a bold, important program. Its launch last night, the attendant public attention, and the fact that the mission's success was dependent on our hardware working have made it once again a primary focus of our company.

From our standpoint, the excitement continues in the weeks after the launch as the telescope comes to life. There are thousands of events that occur as the telescope is commissioned. The timeline that we will be keenly anticipating are the events driven by our mechanisms, including August 3rd when the ACIS instrument cover will be opened ...

Chandra image of a pulsar in the Crab Nebula, spinning 30 times a second, ejecting X-rays over thousands of light years. (Photo courtesy of NASA)

And, on August 3rd, the ACIS cover did open, and our cover motor powered by a paraffin actuator operated flawlessly.

All told, Starsys had more than 15 mechanisms on the Chandra spacecraft, each of them critical to the mission's success. All worked superbly. In the years following the 1999 launch, Chandra has delivered thousands of images, including the one on the previous page that enabled the discovery of a new pulsar in the Crab Nebula: a neutron star spinning 30 times a second, sending out X-rays over thousands of light years, like some maniacal, galactic lighthouse.

There is enormous pride felt by all of us when we see images such as this, as the Chandra launch was a seminal moment for our company, and it validated not only the quality of the devices we were creating but their importance as well. If the hardware we had made at our little company in Boulder had not worked as promised, we would all know significantly less than we do now about our universe.

But the recognition of how close we had come to a company disaster was not lost on any of us. The nature of the paraffin actuators was such that even though our device had acted flawlessly, we had discovered an Achilles heel for our company. If what our motor was attached to did not act as expected, our devices would fail-safe, meaning they would take themselves out of commission before harming something else on the spacecraft. Electric motors didn't have to do that. They could just sit there and hum until whatever was stuck became unstuck. Although our devices had done exactly what they had been asked to do, the spacecraft designers were recognizing that sophisticated mechanisms are better off left to more traditional technologies like electric motors, and that wax was best for the latches. We needed a much bigger tool belt. We were going to have to very quickly figure out how to become a space motor company.

MORE THAN WAX

Artist's rendition of the Chandra spacecraft (Courtesy of NASA)

STS 93 Launch of the Chandra telescope (Courtesy of NASA)

BECOMING A SPACE MOTOR COMPANY

CELL MITOSIS IS A PROCESS of an organism growing by the doubling each of its cells. I remember time-lapse films of this in high school biology: Two cells under a microscope, unaware of the time-lapse camera spying on them, happily vibrating, spinning, and growing in the nutrient medium. Then, at a certain point, they would replicate to four cells, then to eight, and then to 16 on the way to becoming a fully formed organism. I couldn't help but anthropomorphize the moment of division: little cells pausing, realizing they couldn't stretch any further, and then clenching, grunting, and popping from two, to four, to eight, on and on. It looked painful.

In 1997, our company was in a similar situation. We had reached the limit of organic growth[15] as a wax actuator company. To grow any more as a space mechanisms company, we were going to have to provide the real deal: space motors. Wax actuators were simple and the physics were trivial: Confine a material that is trying to expand and harness the expansion to open a latch.

Motors required a deeper, more mysterious magic: harnessing unseen electric fields in just the right way to turn solar-generated electricity into rotation … and then to direct the rotation through gears to a mechanism that can be used to move things on spacecraft—all the while making sure that nothing could fail.

---

[15] There are two kinds of company growth. There is organic growth in which a company's competencies are nurtured and grown into new markets and products; think of Apple selling iPhones first into the U.S. and then Europe and China. Contrasting this is growth by acquisition, in which a company desires to be in a new market, and buys a company already established in that market. Think AOL buying Time-Warner in the late 80s (or maybe best we forget that one).

FROM THE GARAGE TO MARS

We knew enough to know we were not going to be able to figure this out by ourselves. It was too nuanced a device to design from scratch despite the many talents of our engineers. While it was tempting to hire a single, top-notch space motor designer and have this individual develop our space motors, we realized that even if we believed we could pull this off that our customers would recognize the thin veneer of expertise painted over a core of "never done that before."

The only viable path was to acquire space motor expertise lock, stock, and barrel. And the only way to do this was to buy a space motor company—something that couldn't possibly be in the cards for a $6M-a-year nascent space company.

That would have been true except that, serendipitously, a similarly nascent space motor company in Camarillo, Calif., American Technology Consortium (ATC), was having a complementary conversation at about the same time. Founded by a pair of rock-star space motor designers, Jim Sprunck and Doug Peterszcak, they were also stuck, realizing they didn't have the gravitas to take their company to the next level and were wondering if there was a spacecraft mechanisms company out there needing a space motor capability.[16]

The Starsys and ATC conversations converged on May 10, 1998, when Jim Sprunk hunted me down at a space mechanisms conference. He pulled me aside and told me that he was interested in selling their company to us.

This type of thing just doesn't happen. It was as if a stranger came up to me and said, "I've scratched this lottery ticket and won the Powerball ... would it be OK if I split it with you?"[17]

---

[16] Jim and Doug were also struggling with a not uncommon challenge of entrepreneurial relationships. Start-up partnerships are like marriages: two people linked at the hip for the long haul with founder compatibility and complementariness crucial to company success. This is often first tested when success hits. ATC had great early success, but Jim and Doug had diverging ideas of how to take the company through the next steps. With challenging differences between partners, the path forward often results in one partner buying the other out or, in this case, selling the company and dividing the proceeds.

BECOMING A SPACE MOTOR COMPANY

Initially, reasonable thought intervened dismissing Jim's offer as less than legitimate. I knew the company and thought highly of their business, but it just didn't fit into my head that he was suggesting that he would deliver the company to us on a silver platter. It was the antithesis of Negotiation 101. Nevertheless, there was Jim saying, "Want to buy my space motor company?"

It was surreal enough that I took him only half seriously, keeping him somewhat at arm's length by saying I would think about it and get back to him. It took three attempts from him during the conference before he finally said with exasperation on his last try, "Do you have any idea what I'm offering you???" That cut through the rational mind-yell of "too good to be true" that had been masking one of the most significant Nudges of our journey. Jim and I sat down and began to explore the possibilities.

Starsys had no business acquiring a company. We did not have the ability to recover if things went south, and it was likely to not live up to expectations, as most acquisitions don't. We did the math. If we did not book at least an additional million dollars of business in the year following the acquisition, our company would be crippled by the debt. Despite our research and contrary to rational thought, we said "yes."

We followed up with Jim over the next couple months, conducting our due diligence and then moved quickly from concept to terms to final agreement. and a month later, ATC became Starsys Research Space Motor Division.

Sixth months later, we had booked an additional $6M in business.

---

[17] What I did not appreciate until years later, was the investment we had been making in creating an extraordinary organizational culture was having a far broader impact we had expected. Word was getting out about this company, Starsys, and ATC wanted to be a part of the fun. Rather than pursue the dozens of potential aquirers, they were coming to us. Not from a coy position of "We're selling our company. If you're interested, we'll consider an offer." Instead, they were saying "we've chosen you to take our company forward. What price would make this work for you?" A couple of years later, this happened again when a division of a competing company in Durham, NC, split off from the mothership specifically to become a division of Starsys. They all quit the company, then a year later, reached out to us to "bring the band back together."

FROM THE GARAGE TO MARS

We did not realize the dormant hunger in our industry for a space motor company that delivered solutions that *worked*. In our first ten years of business, we had established ourselves as the go-to company for delivering innovative space hardware. When we brought ATC on board, we were a $6M-a-year business that in six months became a $12M+ business. At the point that the ink dried on the acquisition execution documents, we became the company every spacecraft manufacturer wanted to work with and that every space industry enthusiast wanted to work for, as we now had the technological chops to do pretty much anything on a spacecraft that moved.

By becoming the company everyone in the world wanted to work with and for, we had lit a fuse to something that none of us were prepared for.

We were going to need a bigger boat.

SERENDIPITY
A LOOK UNDER THE HOOD ("NUDGES")

YOU COULD SAY THAT WE GOT lucky with ATC, but that would be giving it far too short a shrift.

Serendipity plays an extraordinary, unheralded role in successful entrepreneurial companies. A confluence of events that appears highly improbable conspires to come together to create an out-of-the-blue opportunity that catalyzes wild, unplanned-for successes.

When you hear of these events in a start-up story, they are sometimes treated with a "They got lucky" or similar dismissive comment. That is how it looks but is not how it happens. Familiar adages such as "You create your own luck" or "When opportunity knocks open the door" are on target but give the phenomenon short shrift.[18]

Entrepreneurs radiate a force field of who they are, what they stand for, what they are looking to be, and what they need to become that. Couple this with a deep-seated societal desire for wanting the underdog to succeed, particularly when the mission is admired (e.g., Elon Musk and Tesla) and it creates a broader conspiracy of support than the entrepreneur realizes. The context of "can-do" is picked up by those around and broadcast as if by coconut telegraph: "Did you hear about this cool company Starsys ... they push fun as a value and make flaw-free space hardware. How great is that!?"

---

[18] If you want to create an animated conversation with one or more successful entrepreneurs, simply ask them about their take on serendipity, the role it played in their company's successes, and if there was something bigger than dumb luck at play.

FROM THE GARAGE TO MARS

The gun is cocked for random connections to forward the success. It becomes a matter then of recognizing these magical intersections as they occur and running with them. The entrepreneur's intuitive sense of "This is something I need to follow up on" becomes highly tuned. Time passes and confidence builds in acting on these hunches—to the point that the entrepreneur notices the "tells" (signs) of an exceptional opportunity and learns to pivot intuitively and quickly when they show. As more time passes, it appears in retrospect to be an extraordinary ability to create one's own luck.

These magical intersections may be manifested in the person seated next to you at a dinner ... or in a networking introduction at an event ... or in any number of dimensions somehow resulting from the subconscious intuitive thrumming that entrepreneurs become attuned to sensing and trusting—the feeling that heading

---

Tibbitts Tip
Serendipity—The 7 Habits

There is much that comes into play when it comes to luck—the things that conspire such that good fortune happens "out of the blue." Starsys was built on what some might call good fortune. We *were* a charmed company. But by design, not accident. With a nod of the hat to Steven Covey, here are the seven habits we found create serendipity:

(1) **Vision:** An "all-in" commitment to a worthy goal

(2) **Impact:** A cause and mission that matters to more than yourself

(3) **Giving:** A desire to pay it forward; to help those around you with no expectation of return

(4) **Humility:** Operating with humble genuineness

(5) **Caring:** A desire to connect with people at a meaningful level. Deep interest in those you meet. and hearing their story.

(6) **Belief:** The certainty that the universe can be miraculous and amazing things can and do happen

(7) **Worth:** Owning that you are worthy of good fortune.

SERENDIPITY / NUDGES

down a particular road less travelled is likely to lead to something of great import.

The entrepreneur self-talk that goes with this process might sound like "I'm not quite sure why but I think I need to go to that event" or "There is a reason I am sitting next to this person at dinner, and I need to figure out what it is."

The other element that is key to leveraging serendipity is an unshakeable belief that these moments do happen, can be recognized as such, are deserved,[19] and need to be grabbed and run with. This effect brings to mind a quote attributed to the German philosopher Wolfgang van Goethe:

> *"Concerning all acts of initiative (and creation), there is one elementary truth that ignorance of which kills countless ideas and splendid plans: that the moment one definitely commits oneself, then Providence moves too. All sorts of things occur to help one that would never otherwise have occurred. A whole stream of events issues from the decision, raising in one's favor all manner of unforeseen incidents and meetings and material assistance, which no man could have dreamed would have come his way. Whatever you can do, or dream you can do, begin it. Boldness has genius, power, and magic in it. Begin it now."*

There is often a subtle physical response to these events that becomes the entrepreneur's tell that serendipity is afoot. When a life-changing intersection suddenly materializes, I will often get a flush of goosebumps that whispers, "It is happening again." I have learned that it is best to keep the tell to oneself. I've learned from experience that saying "Ooooo! I just got goosebumps!" in the middle of a strategic business meeting mucks with the workings of good fortune.

This is our intuition nudging us to take the chance, open the door, look around the corner, peer under the rock. Powerful mojo

---

[19] A doff of the hat and thank you to Carl Jung.

comes when we learn to identify our intuitive nudges and act on them.

An example of one of these intersections was the invitation that I received to a Space Shuttle launch from Boeing, the prime contractor, as an acknowledgement for our company's contributions to the International Space Station program. I didn't have time to be away for several days in Florida, but I had an inkling that there was high purpose at stake here, as it was an opportunity for random cross-pollination with lots of space folks from different backgrounds, all squished together into the close quarters of a Cocoa Beach field trip.

After getting on a bus for a ride to the launch site, a fellow looking for a seat saw my name tag, had heard we were an interesting company, and asked if he could sit next to me. During the 15-minute ride to the launch pad, he made an offhand comment unrelated to our business that his company had a division that made batteries for spacecraft and was having a problem with battery cells that went bad. He further explained that when a battery goes bad it gets hotter than the remaining batteries.

I then floated a potentially harebrained idea: If the battery temperature increase triggered a wax actuator that flipped a switch

---

Tibbitts Tip
Trust Your Nudges

Learn to hear and trust your intuition. The best way to do this is ... trust when your intuition nudges you. We all have physical or emotional signs when we are aligned with our intuition. Learn these whispers from when your intuition gets it right. For me, its goosebumps or an unexpected emotional rise. Find your own tells. When serendipity happens, lessen "It can't be that easy...", "That's unreasonable...", "I can't pull that off..." Replace with "But what if...?", "I would always regret not knowing..." "What's the worst that could happen?" The more you trust your intuitive nudges, the more they *can* be trusted.

SERENDIPITY / NUDGES

taking the suspect battery out of the circuit, could that be a solution? He didn't know but gave me the phone number for someone who would. As he handed me the scrap of paper, I kept to myself the flush of goose flesh.

Back in Boulder, I considered not making the call, as I was sure it was too long a shot. But over the years, I had seen too many of these magical intersections work out not to pursue it a bit further. I made the call and three years later, Starsys was the world leader in battery bypass switches for the spacecraft industry, regularly picking up multi-million-dollar orders for wax-powered battery bypass switches. Had I boarded a different bus, we would not have gotten into that business.

I've come to call these moments *NUDGES*—an "aha" in which a seemingly random connection with the promise of being life-changing appears out of nowhere. A Nudge is always followed by rational thought, declaring, "No way. It can't be that easy," followed by intuitive thought suggesting, "… but just maybe it is." I've learned to look for a serendipity triad and run with it when it happens:

(1) A seemingly serendipitous intersection of events, people or thought, followed by …

(2) "No way. It can't be that easy," rational thought, followed by …

(3) "Yeah, but maybe this is an opportunity not to be missed" intuitive thought.

After decades of experiencing these wonderful moments, I'm finding now that the first step occurs more frequently; the second step appears as a whispered speed bump, with the final step being more akin to, "It's happened again! Let's go!"

FROM THE GARAGE TO MARS

In the childhood game *Chutes and Ladders*, now and then you got lucky and landed on that one spot that took you straight to the top of the game. I loved the feeling when it happened then and love it just as much now. It's a simple reminder that we really do all live charmed lives if we are open to these amazing, life-altering Nudges.

ON MARS

OUR ACQUISITION OF AMERICAN Technology Consortium (ATC) was like strapping a JATO rocket[20] to a VW bug. You knew it was going to go wicked-fast, but had no idea in which direction.

To the amazement of all, our heretical ideas about what a space company culture should be had led to us becoming the go-to company in the space mechanisms community (at least the go-to company if you wanted wax on your spacecraft).

We were unaware how much our having only this one arrow in our quiver had been moderating our growth. We were blindsided when the ATC acquisition removed the shackles by giving us the technological chops to deliver pretty much anything that moved on a spacecraft. For the first year following our acquisition of ATC, we had held our breath, knowing that we needed $1M in new business to result from the merger for us to not regret the deal. We finger-crossed that this might be as much as $2M. By year-end, as I've mentioned, we had booked $6M.

It is a transition few entrepreneurs are prepared for, when a decade of doing anything you can to talk customers into buying your products, morphs overnight to the very different problem of having more work than you can easily handle. We had not expected nor prepared for the rocket sled ride of success that came with the acquisition of ATC.

---

[20] Jet Assisted Takeoff Rocket used by the military to accelerate planes quickly to high speeds for take-off on too-short runways. The analogy may seem far-fetched, but some knucklehead actually did this back in the 90s. It had unintended consequences.

FROM THE GARAGE TO MARS

But oh, what a glorious problem to have! Imagine walking across a desert for years, eating the occasional prickly pear cactus and horned lizard you come across, while you dream of hamburgers and root beer floats, and then you come over a dune and see a banquet table laid out for you with everything you dreamed of and more. *"Holy smokes! Is that a filet mignon with Béarnaise sauce and a box of Krispy Kreme donuts?!"* It takes a while to morph your entrepreneurial mindset from handling starvation to managing gluttony.

Nothing embodied this new reality better than our once-in-a-lifetime chance to go to Mars.

NASA can only go to Mars once every two years, as with Mars circling the sun about once every couple of years, our two planets line up only every 26 months.[21] The mission that was planned for 2003 was ambitious and built on the success of the 1997 Mars Pathfinder, which had become a monster technological and public relations success. Pathfinder's purpose had not been groundbreaking Martian science. Rather, it was a demonstration of the feasibility of radical new technologies such as air-bag landings and robot rovers. With the Pathfinder success and public support, NASA was able to convince Congress to fund an ambitious program to land two rovers on the Martian surface in 2004 that would bristle with scientific experiments, focused on answering the Big Question of whether or not there had once been water on Mars. The shared thinking was that if there was once water on Mars, there was once life of on Mars. There was also an unspoken hope that just maybe one of the cameras might send back a picture of a fossil, changing forever our view of the universe.

Two landers were planned to provide what NASA calls redundancy. The chances were not good for the success of a single

---

[21] A major reason that sending people to Mars is crazy-hard. Once you get there you must hang out for two years before you can come back, and a major worry of NASA is the psychology of that level of isolation from the rest of the human race. Astronauts make poor hermits and vice versa .

mission. By having two, staggered as to when they would arrive on the Red Planet, the odds that the first would fail would be mitigated on the off-chance that they could correct problems mid-flight on the second, bettering the odds. NASA administered an essay contest amongst school students to name the rovers and to further fuel public interest in the program, eventually settling on *Spirit* (the first) and *Opportunity* (the follow-on).

The mission was a spacecraft mechanism designer's wet dream: prior to launch, the rovers were to be folded up like a contortionist in an acrylic box, surrounded by un-deployed airbags, tucked into the spacecraft nosecone, and sent to Mars, where eight months later, they would scream into the Martian atmosphere at 12,000 mph, slow to the point a parachute would open and inflate its airbags. Then, 100 feet above the Martian surface, separate from the parachute, bounce across the Martian surface until it came to rest, deflate the airbags, retract them into the lander, unfold like origami, and finally, if the above all worked as planned, unfold its wheels, solar panels, and antennas, and drive off the landing platform to explore the Martian surface. What could possibly go wrong?

With the risks high, JPL chose to do much of the work in-house, including designing the motors that would power the unfolding choreography, and those that would drive the rover around Mars. There were few elements as critical as these motors—any one of the dozen motors failing would mean mission failure.

In 2000, Starsys had suggested that we could do the work for NASA, building the gearboxes and attaching them to small electric motors made by another company. JPL had politely declined, keeping that work with their in-house team. In the year following that decision, we heard the occasional rumor that things were not going as well as expected on the program, and in the summer of 2001, I got a call from one of our champions within JPL who had just left a high-level program meeting that afternoon.

"Scott, I just left a NASA Program meeting. Mission leadership is worried that the in-house effort might be a schedule risk. As the launch window is narrow and a schedule slip would mean a two-year delay, funding has been authorized to start a parallel effort."

"Are we in the running?"

"Yup."

"Who else?"

"No one. You're it. Program thinks Starsys is the only company that can succeed in the timeframe. It's to be a sole-source contract.[22] It's yours if you want it. It's coming to you soon and it's a big deal. You're being brought in to save the mission schedule. You'll get a call tomorrow asking if you're interested. Act at least a little surprised, but I wanted to make sure you were ready for the call. We will need 27 different actuators for each of the two rovers."

We did not realize it at the time, but to give this much work to a single supplier our size was unprecedented. Typically, this type of work was split between multiple vendors to ensure that each would not fail from having too much to do. In this case, the situation of being far behind schedule warranted the Hail Mary approach of picking one supplier for it all, as there was not the time for multiple competitive contracts.

"Scott, JPL's betting the mission on you guys. Don't screw it up."

An entrepreneurial company's arc of growth is rife with karmic balance. The occasional horrific, bring-you-to-your knees bad luck balanced by the occasional too-good-to-be-true, opportunity-knocks moments. You've got to grab the latter by the throat and believe it is deserved to have a chance of evening the karma scales.

---

[22] Sole-source contracts are given to NASA contractors, when it is in the best interest of the program to not compete the contract, but to provide it to the one contractor that the Program believes is the only acceptable supplier for the task. Contracting then becomes focused on the pricing being fair to both parties rather than a competitive bid process.

ON MARS

We did just that, and by the time the call came, we had convinced ourselves we were the team to save NASA's bacon. When the call came, the NASA project engineer did his best to keep the urgency out of his voice: "Starsys, we need you to provide a quote for us ..."

Knowing that we were the only option, we made sure the price matched the importance of what we were providing. We needed the financial resources to pull out all the stops and pull this off.

Two weeks later, we were under contract for a multi-million-dollar parallel effort. Four months later, JPL cancelled its faltering in-house effort, and we were it. The mission truly did now rest on our shoulders. All we had to do was design, develop, and deliver more than 100 flawless space motors in nine months (including spares). We should have been frightened but were not. If NASA thought we could do it, who were we to disagree?

The NASA Program fired up the company. Much work done in space is not easily explained to your neighbor, much less understood, but this was different. Everyone could get excited about driving Rovers around Mars. JPL sent us a CGI video of the mission, made to the soundtrack of Lenny Kravitz' "Fly Away," with the direction to crank the volume whenever we played it.[23]

> *"Let's go and see the stars, the Milky Way or even Mars, where it just could be ours."* [24]

CGI was new at that time. To actually see what the things we were building would look like in operation, was impressive. Overlay that with 120 dB power chords and lyrics about going to the planets, accompany with air guitar, and you could almost believe that making space hardware was bad-ass. We were stoked!

---

[23] It turned out NASA had not asked Mr. Kravitz' permission for NASA to use the soundtrack, and shortly after we received the video, an email was sent asking us to please erase the file, which we conveniently ignored, the bootleg nature of the clip adding to the edginess.

[24] Copyright 1998, Lenny Kravitz Productions Inc

FROM THE GARAGE TO MARS

The program received much attention from NASA, with multiple meetings attended by a dozen people including Mission project manager Peter Theisinger—a good-humored fellow, reminiscent of a skinnier version of Santa Claus with rosy cheeks and an easy laugh. We found out later that on his first visit to the company, he stopped after he got out of his car, paused to look at our building, and said to himself, "If these people don't do their job just right, we will fail," before walking in with the "I'm sure you can do it" bravado we needed to hear from him.

To our (and NASA's) surprise, we did the job with little drama. The program went smoothly, and we delivered the motors on time. NASA assembled the spacecraft and on June 10th, 2003, *Spirit* was launched from Cape Canaveral. Its twin, *Opportunity*, launched three weeks after that.

Eight months later, on January 4th, 2004, the *Spirit* Rover arrived at the Red Planet. To celebrate the landing, we catered a party at a local hotel, the Boulder Outlook, with champagne, snacks, and a live CNN feed of the control room. A Mars landing does not culminate in an impressive visceral event like those of a splashdown or moon landing. Mission success is indicated by nothing more than light turning green on a console indicating touchdown, or a first transmission of a beep from the surface that all is OK.

About one hundred of us anxiously watched CNN as the news anchors reported on the various steps in the landing process and what was happening at that particular time. My then-11-year-old son Ryan was also there with us to share in something that would hopefully be history-making for our company.

The period of time prior to the landing was officially called by NASA "Seven Minutes of Hell," as the spacecraft had to slow from 12,000 miles per hour and do hundreds of things just right to make it safely to the surface. Despite NASA's confidence, when you pulled them aside and asked what they really thought the odds were, they were worried. There were many fingers crossed and probably a few chickens sacrificed in Pasadena that night, as all hoped against the odds that NASA would pull it off.

The coronal discharge during reentry (essentially the aurora borealis that forms around the spacecraft as it plunges into the atmosphere) meant that communication was limited to a few beeps, such as "heat shield released" or "parachute deployed," each relayed by CNN and celebrated by us with a holler and a toast.

And then the TV went blank.

NASA was unprepared for the national attention focused on the landing and the sheer number of people who would be streaming the CNN feed. As the landing approached and the nation got online to watch, the servers finally crashed. As NASA scrambled to get additional servers online, we were left not knowing whether the Rover was now safely on the surface or a crumpled heap lying in the middle of a self-made crater.

After ten minutes of not knowing, the CNN news feed came back online just in time for us to see a control room full of NASA scientists watching screens and not moving a muscle, waiting for the "beep" that indicated the spacecraft had landed safely on Mars.

The beep came, and the control room exploded. The anchor announced, "*Spirit* is now safely on the surface of Mars and sending signals, indicating all systems are functioning properly." We jumped up and down, wept with joy, and hugged each other indiscriminately. We all stuck around to see the press conference summarizing the events, and there was Pete, on the stage, with dozens of cameras pointed at him, with the smile of a man who had wrongly assumed a couple of minutes ago that he would be leading off the press conference with "We are, of course, disappointed by this result, however ..."

FROM THE GARAGE TO MARS

Instead, he let the applause build and eventually die down, and through a smile that wouldn't stop, led off the press conference with "I've told this story many times. A couple of years back, I remember we were having trouble getting parts built. A company in Boulder, Colorado, picked up the challenge. I went out to the company and stood in the parking lot and said to myself, "*If this company doesn't do its job just right, we will fail.* They and similar companies deserve our heartfelt thanks."

We didn't hear the rest of his comments as we exploded in hoots and hollers of "That's us!"

I don't remember much of the next half hour. Although there was not much on CNN anymore other than replays of the "Seven Minutes of Hell" video, it took a while before we reluctantly headed home. Ryan and I were last, walking to the car and not saying much while we headed north for the 15-minute drive to our home in Niwot. Ryan understood the success, but not the depth of importance. For him, it was a cool thing that Dad had done in his space company. For me, it was much more. We were quiet on the drive home with the occasional "That was really, really cool" comment.

We got home at 11 and Ryan headed to bed. I wasn't ready to let this feeling end. I was basking in the rosy glow of this particular success, where I didn't want to leave the self-talk of "I can't believe we pulled this off!" I knew that tomorrow would not have the same edge.

On a whim, I headed to the study. With the lights low, I settled into the leather chair, in front of our big cherry desk, the slight scent of Pledge adding to the warmth of the room. I turned on the computer and clicked on the JPL website. As the screen filled, a picture appeared that gave me goosebumps. There, in high resolution, was the first picture back from the rover.

ON MARS

I froze, awed, and stared at the picture, taking it all in. There were likely tens of thousands of people just like me, scattered around the globe, looking at the same picture, sent only minutes before, experiencing something as a body of people that no one in history had ever seen before. NASA has done an amazing job of bringing us along to share in the wonder, letting us experience major discoveries simultaneously with the explorers. I had chills thinking that this must have been what Galileo thought when he was first to see mountains on the moon through his telescope. We now get to go along for the ride, sitting in the crow's nest as a new land appears on the horizon. But my experience had a special richness—I was looking at a picture that our company had helped create.

I zoomed in to look at the details of the planet, and then zoomed back to take in the panorama. The color and resolution were jaw-dropping, giving me the feeling I was there. That's when I saw a smudge on a flange in the middle of the camera tower that had not yet been deployed and was lying on top of the deck. I centered on the smudge, and *click, click, clicked* to zoom in, saying out loud, "No way ... there is no way ...," and then stopping when I could see the details of the mark. Staring. Motionless.

It was the Starsys logo, clear as day, etched into the camera tower flange. The only logo—ever—of a company like ours that had been shown on another planet. The best product placement ... EVER! I spent minutes zooming in and out, and then was suddenly flooded with emotion as I flashed back to walking through the JPL campus and saying to myself, "I would give anything to play a part in this." It had come true. That *NUDGE* I had had driving into Boulder. It was now real in every sense of the word. Our company had made it possible to get to Mars, and as a reward, this logo had ended up in just the right place as a reminder that what you sometimes imagine, really can come to be.

FROM THE GARAGE TO MARS

The Starsys logo on the Mars rover (Courtesy of NASA)

After another 15 minutes, I turned off the computer and went to bed. The next day, I called a company meeting. I had taken a screenshot of the picture. I told the team, "You aren't going to believe this" and projected the image on the wall. I asked if anyone knew how this got there. A hand rose in the back.

"I put it there."

Kyle Hickey was a young, sharp, rising-star designer.

Everyone turned and stared.

"Yeah, when it came time to finish the drawings for the motor on the camera tower, I thought it would be great to sneak our logo somewhere on the hardware, so I looked up the drawings for the Lander and where the camera was going to take the first picture. So I put it on that part, oriented to be in the field of view. It worked!"

The others in the room continued to stare.

"I wanted to surprise you guys. Pretty cool, huh?"

The company gave Kyle a standing ovation.

* * * * *

I doubt that we ever got an order for spacecraft hardware from that product placement, but that was one of the coolest things our team ever did. Not just for the cleverness and foresight that went into it, but for the sentiment behind the action.

The company we created—this organization that was constantly looking for ways to give back to the employees who were giving so much to us—had created a body of employees who were looking for ways to give back to the family. The logo on Mars was simply a surprise gift from Kyle to the rest of us.

There is no way as a leader I could have managed to create an outcome like that: "OK, now, I want us to find a place to put our logo on Mars." But by being committed to Starsys having a company DNA that held *Family* above all else, in response, Kyle had created this extraordinary thing that for years has reminded me of the magic that will happen when you honor the people who are honoring you by helping bring your entrepreneurial vision to life.

* * * * *

The expected mission life of the rovers at the time of the Mars landing was 90 Mars Sols (days), or about three Earth months. One of the reasons for the short life expectancy was the concern that the actuators Starsys assembled might not last beyond that in the Martian carbon-dioxide atmosphere. Our motors exceeded expectations: The *Spirit* rover continued to explore Mars for six years. The *Opportunity* rover kept on functioning perfectly for an astonishing 14 years, 52 times longer than the "warranty period!"

A 360-degree composite panorama of Greeley Haven.
Taken by the *Opportunity* Rover December 2011. (Courtesy of NASA), .

The rover tracks show the route *Opportunity* followed as it explored various areas of interest. 16 Starsys actuators are shown on various MER mechanisms.

COOL FACTS FROM THE MARS EXPLORATION ROVERS

A major concern with respect to lifetime was Martian dust. Frequent Martian dust storms cover everything in a fine, red dust. NASA was certain that the accumulated dust would cover the solar panels, robbing the rovers of power and ending the mission. The dust storms did lessen power to the rovers, but in a strike of cosmic good fortune, Mars also experiences frequent "dust devils," small,

Mars "dust devil" (Courtesy of NASA)

wind-powered micro-tornadoes that the rovers would occasionally encounter, conveniently blowing the dust off the solar panels, and returning them to full power.

* * * * *

When the first pictures came back from the *Spirit*'s sister rover, *Opportunity*, the control room was gobsmacked. Against all odds, *Opportunity* had rolled into a small crater in the middle of a Martian desert, a cosmic "hole in one." It stopped right next to a shelf of bedrock that had been exposed by the wind. It was the first bedrock ever seen on a planet other than Earth and one of the first areas *Opportunity* explored.

Opportunity subsequently drove to the bedrock to take a closer look. Cameras showed that small nodules, blue in color, were scattered amongst the rock. JPL called these "blueberries" because of their grayish tint against the red rock. Analysis showed the presence of salt, which can only be formed by water, and provided the first positive evidence that Mars had once been awash in water.

Image of "blueberries" (salt nodules) on the bedrock near where the rover landed. The circular mark show where *Opportunity*'s grinder tool exposed the rock for analysis.
(Courtesy of NASA)

FROM THE GARAGE TO MARS

This discovery has fueled multiple missions since; if there was water, there likely was (or is) life.

* * * * *

A fact not well known is that there are already Mars rocks on Earth. Over the past several hundred million years, the occasional asteroid or comet would impact Mars, creating the craters we see on the surface. The resulting explosions hurled thousands of tons of Martian rock into space, most of which landed back on the planet, but some was put into orbits that sent chunks to Earth, where they became meteors, many of which made it to the ground.

If you are looking for Mars meteors, the place to go is Antarctica, as any rock on the surface of the ice field is likely a meteor. And some of those have been confirmed as being from Mars by the presence of carbon dioxide in the rock. While *Opportunity* was cruising across a Martian desert, it encountered a rock in the middle of nowhere, all by itself. The rock came to be called "Bounce Rock" because *Opportunity* bumped over it as it inadvertently drove over it.

When *Opportunity* encountered this rock, it analyzed its mineral content. The analysis was then compared with analyses of various Antarctic Mars meteorites, and a perfect match was found. The rock *Opportunity* encountered was a "sister rock," both blasted from the ground during the same event.

* * * * *

ON MARS

"Bounce Rock" (circular area showing where the grinder tool exposed rock for analysis.
(Courtesy of NASA)

LOSING KURT

IT WAS 2001 AND WE WERE deep into the first years of our rocket-sled ride of entrepreneurial growth. After acquiring American Technology Consortium (ATC), we faced the heady experience of receiving calls from the companies and organizations we had for so many years longed to work for—NASA, Lockheed, Boeing, European space companies. They were now knocking on our doors, asking us to make devices that five years ago we could not have imagined, much less known how to build. It was a frighteningly dynamic time, with multiple complex products being built and insurmountable problems showing up weekly that had to be surmounted.

Leading our company felt like piloting a spacecraft at blinding speeds through a canyon with narrow, unforgiving walls. All fine, if you kept everything pointed just so, but a swerve at just the wrong time could result in disaster. The key to keeping it all together was our program managers (PMs), who were committed to delivering products on time, to spec, and below cost.[25]

Kurt Lankford was a rock-solid program manager in charge of Radarsat, one of our bigger programs. It was a sophisticated electronics control box and motor that was to control an antenna on a major Canadian Space Agency spacecraft that would map the earth with radar. Radarsat was also outfitted with technology that was very new to us: space electronics.

---

[25] Flying in the face of conventional wisdom best expressed in a T-shirt popular at the time with engineers: A triangle with "Good," "Fast," and "Cheap" at each apex and the headline below "Pick Any Two."

FROM THE GARAGE TO MARS

The rules for making space electronics are a byzantine maze of Thou Shalt Nots, strangely reminiscent of the troll guarding the bridge in *Monty Python and the Holy Grail*, "What is the airspeed velocity of an unladen swallow?" Get one thing wrong—no matter how small—and you were thrown over the bridge.[26]

Space electronics problems are never small. They break your heart and can bring companies to their knees.

But we had Kurt on the program. Tenacious as a badger, quiet as a koala. I hired him on a recommendation from his former CEO that I put much weight on: "I would rehire Kurt in a heartbeat. If you can hire him, do so. You will not regret it."

Kurt was somewhat of a lone ranger. He backcountry skied and had written a seminal book on Colorado backcountry skiing. He was a great father with a great wife, Karla, and two young children, Austin and Ashley. His desk was swamped with artwork drawn by his children, which created a cocoon of family around him while he worked on the hard problems.

> **Tibbitts Tip**
> **Recommendations from References**
>
> Even great employment interview techniques aren't 100% effective; we found that we were at best only 75% effective with our interviewing.) A recommendation along the lines of "If you have the opportunity, hire Monica. You won't regret it." Or "Hire him even if the role isn't a perfect fit. You're sure to find a role in which he excels." These statements will not come from HR, which will be politically-correct-bland, but from their manager, or better yet, the CEO. With that kind of recommendation, we were close to 100% success with our hires. Kurt was my first experience with this.

---

[26] Imagine you need a specific electronic memory chip in your control box, and it takes six months and $50,000 to order the minimum 100 units, of which you will only use one. No problem so far, as this is something the customer pays for. Just before it arrives, NASA reports that on Mars, a duplicate of the chip you are about to use had a cosmic ray hit, which spontaneously shut down the spacecraft. As a result, you can't use the chips that just landed on your dock. It turns out you can go to Radio Shack and buy a replacement part for $35, which will probably work in space ... but Rules say it must be fully tested for months in a radiation environment. Although you know it will pass, testing takes an additional year and $250,000, but those are the Rules. Dang. You just lost $500,000 dollars and 1.5 years of schedule.

LOSING KURT

He was also a mountain runner and ridiculously fit. He often used his lunch hour to run up mountain trails.

Despite his quietness, though, he was a closet rabble-rouser and rogue.

This tendency was in full display one cold, sunny, snow-covered day in January when Kurt coaxed a dozen of our engineers outside during lunch break with a bungee cord in one hand and a small plastic sled in the other. Earlier, the company had bought 200 feet of military-quality bungee cord with a loose sense of "we are engineers and will find something fun to do with this," and that day, Kurt had found the "something fun."

Much to the group's glee, Kurt had built a jump on the field in front of our building. He suggested we double up the bungee into a slingshot and have one engineer sit on the sled holding the middle of the bungee with 10 engineers on the other two ends of the bungee, encouraging them to walk 100 feet to stretch the bungee to its absolutely limit just short of snapping. Then, with their muscles quivering, he yelled, "Release the engineer!" This resulted in catapulting a 130-

> **Tibbitts Tip**
> **The Power of Play**
>
> We did not at first recognize the power of play, we thought of it initially as a fun way to spend time during work breaks. It became far more than that, as it brought the team together, created trust, reminded the team that work was not everything, and spawned creativity. Particularly when the fun was innovative and edgy. Different than putting a foosball table in the break room, our playtime included launching engineers with bungee cords, playing the occasional Road Runner cartoons in the break room (complete wth buttered popcorn), competitions with mouse-trap powered race cars and parking lot roller-blade hockey. The play breaks more than paid for themselves in company efficiency and loyalty.

pound engineer at 30 mph in about 2.5 seconds just before hitting the jump. The landings were not pretty.

That was Kurt. He was under incredible stress with Radarsat and as the program leader, he had to sort it all out. He used the lunch-break runs to relieve the stress and sort through the problems.

* * * * *

At 1 a.m., the phone rang, waking my wife and me. Karla explained that Kurt had not come home from work that night. She was worried, alternating between "This is silly. I'm getting all worked about nothing," and a legitimate, serious concern that something horrible might have happened. She wanted to know if I had any idea where he might be, hoping against hope that I had sent him on a business trip.

I told Karla I would call her back and then called several other folks in the company. No one had seen Kurt since lunch, and he had missed some important afternoon meetings. Our operations manager, Doug Monick, volunteered to drive to the company to see if there were any indications of where he had gone. Doug drove back to work and checked Kurt's email and the sign-out sheet at the front desk. It looked as if Kurt had left for lunch but then had not returned. Something wasn't right.

It was now 2 a.m. There wasn't much more I could do before morning, but I decided that doing anything was better than just waiting. I hopped in my car and joined Karla driving around the area looking for Kurt's car at various trailhead parking lots. We checked in with each other every five minutes or so by cell phone. Neither of us could come up with a scenario that wasn't bad news. We nervously joked that maybe he had run off to Mexico with some woman, but knew that we were grasping at straws. We eventually decided to stop searching and get some sleep.

LOSING KURT

First thing the next morning, I convened an all-hands meeting. I explained that Kurt was missing and that it did not look good. I went to his office and sat in his chair. What struck me was the freshness of it. If Kurt was gone, his office didn't know it yet. It was like an expectant puppy waiting for his master to get home.

The company settled into the day, everyone quietly hoping and praying that this would turn out all right. Karla dropped off Austin, her four-year-old, at Starsys before the police arrived at her house to get involved with the search. We entertained him while a couple of us waited for the call in the conference room. This small group included Jenny Donaldson, our HR manager, who had been through things like this before and was preparing us all for the worst.

The call came in at 9 a.m. that Kurt's car had been found at a trailhead of Mt. Sanitas. The trailhead was a 20-minute drive from our company and adjacent to a parking lot that we had not checked the previous night. There was hope—he could have stumbled off the trail, been hurt, and then lain there overnight waiting for help. It is strange how quickly we lean that way, when we get a small piece of information that indicates "it could have worked out OK." That was not to be.

My memory of the next few minutes is vivid. With the leadership team around the conference table, another call came from Karla, choking out words while struggling to keep it together for her son and daughter. "Kurt's gone," she said. "What am I going to do?"

The floor dropped out from us all. Quietly, under my breath, I said, "F***."

While running up Mt. Sanitas at lunch time, a 1,000-foot vertical climb trail on the edge of Boulder, Kurt had experienced a fatal heart attack. He had collapsed a bit off the main trail, likely after stepping off the trail to take in the view of Boulder below. I

FROM THE GARAGE TO MARS

> **Tibbitts Tip**
> **Dealing with Great Loss**
>
> Dealing with great loss or death of an employee is hard, but a powerful "leadership moment." There are lots of ways to do this wrong, and a few to do it right. I found in this case (as well as dealing with 9/11 as the company gathered and saw the second plane hit the tower), I needed to be honest with the emotions I was experiencing, particularly as they were representative of those in the company. As I dealt with both of these, I let the tears and emotion I was feeling manifest when I spoke to the company, as long as it didn't prevent me from communicating. I think both Reagan and Bush did this well with the loss of *Challenger* and 9/11. They both honestly conveyed the emotion the country shared.

looked at the people in the room with me, and then the tears came. We turned to Kurt's son, Austin, and without leading on as to why, suggested he go home. While Jenny took Austin home, I announced another all-hands meeting.

I stood in front of the room, not knowing what I was going to say. I had always known that at some point, I would have to deal with a death at our company. Since many of our folks were involved in risky activities like climbing or motorcycle riding, I expected something like that would be the cause—not a heart attack. But here it was, and what was I supposed to say and do? It was not so much hard as it was a matter of simply being honest.

I spoke from the hurt in my heart, and from my own experience as a father. It was a struggle to get the words out through the tears. Starsys was a close family of 100 people at the time.

I asked our people to keep Kurt's family in their prayers, and that they do what they felt they needed to do during the day—whether that was going home, or working, or just talking. Afterward, everyone headed back to their offices and the company settled in to mourning for Kurt.

That day, people started to ask what they could do to help the family. One of the engineers offered his home, and about 20 people

LOSING KURT

got together that night to talk. Over the next couple days, we set up a fund that our people could contribute to, and we also made it possible to provide support in the form of contributing unused vacation; Starsys would then translate this support into equivalent dollars and make a matching contribution. The contributions were astonishing. Several people offered up to a week of vacation. Spontaneous and generous contributions also came in from the companies Kurt had worked for previously. We were ultimately able to provide a check for $50,000 to help with Ashley's and Austin's education.

> **Tibbitts Tip**
> **Sharing in the Giving**
>
> Sharing in the giving was a powerful way for us to be a family. Although the team could have afforded to pay $50K to Karla and her kids, by allowing the team to also give, it gave them a way to personally be involved in the healing, and was a reminder that we were family and sharing in the good and the bad.

Other forms of support came in as well—such as the letter that Michael Hecht, a NASA JPL program manager who had worked with Kurt on a previous Mars Project, wrote to Kurt's children describing all the special things their father had done for the Mars Program. In addition to the letter, he sent a package full of Mars memorabilia.

* * * * *

Our people continued to grieve throughout the month. We left Kurt's desk space untouched for a while. Then when the time was right, Karla came in quietly with the children after hours, packed up Kurt's belongings, and took them home. As it turned out, Kurt had been quiet about his accomplishments at work, and the family had little knowledge of the details of what he actually worked on. We wrote the children a letter that described all that he had done, and the things that he contributed to our company and the space program.

FROM THE GARAGE TO MARS

Two weeks later, we had an idea. The Radarsat II Electronics Control Unit (ECU) Kurt had been designing had some space on it where a small message could possibly be etched. We asked Karla if her children would like to put a message on the cover that would be going up to the heavens to circle the Earth. When Karla said yes, we called Gary Heinemann, the project manager on Radarsat II, and asked if we could put a small message on the box. The call came back the next day. Gary and his team had checked with MacDonald Dettwiler, the ultimate customer, which was based in Canada. Gary's words were emphatic: *"You put a message on there and make it as big as you can!"*

Karla, both of her children, and Kurt's mother each wrote a personal message, which were then etched onto the cover of the ECU, visible on the spacecraft for all to see. I expect there were more than a few technicians and engineers at the Canadian Space Agency who, while readying the spacecraft for launch, would have seen the cover, paused, understood, and said a quiet prayer.

A month after Kurt's death, Karla asked if the family could come to Starsys to say thank you. With everyone packed in the meeting room and quiet sniffles heard from most, Karla talked about how hard the month had been, how thankful she was for her Starsys family, and for all that had been done.

She then explained that the children had a gift for us. Over the next ten minutes, they quietly walked up to each of us, said thank you, and handed us an Origami crane they had made, the crane being a symbol of hope and healing during challenging times.

It was an emotional closure to a very rough month. Just about every office in the company kept one of the cranes in some special place as a reminder of Kurt.

After the loss, quite a few people made a point of stopping me and expressing a similar version of the same basic message: "I never fully understood what people meant when they said Starsys was a family. Now I do."

LOSING KURT

Kurt Lankford (center) holding a version of the heat switch he built for the Mars rovers, *Spirit* and *Opportunity*.

Origami crane made by Kurt's children.

FROM THE GARAGE TO MARS

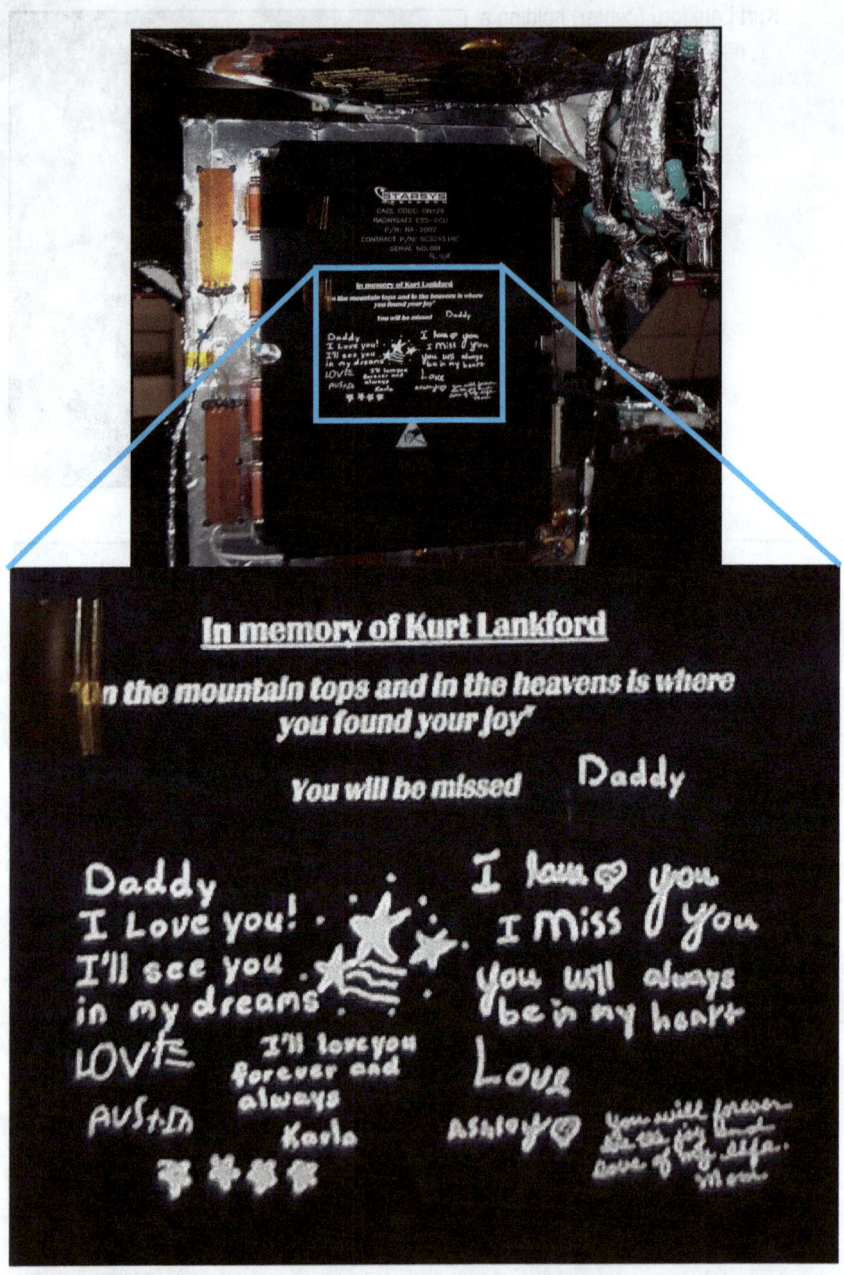

On December 14th, 2007, Radarsat II launched from the Baikonur Cosmodrome and continues to capture images of the Earth with Kurt's electronics pointing the antennae that transmits the images to our planet. (Photo courtesy of McDonnel Dettweiler)

THE MORNING MEETING

Starsys Will Be the Best Spacecraft Mechanisms Company in the World by Being One of the Best Small Companies in the World.

IT WAS AN AUDACIOUS MISSION statement, initially floated as a whisper rather than a declaration. But as we slam-dunked one mechanism program after another, we developed the courage to put it on the wall in big letters. We were all gunning for Starsys to be one of those companies you hear of for their great company culture: Apple, Google, Southwest Airlines ... but Starsys? Like those signs you see for the locations of exclusive restaurants: London, Tokyo, New York ... Moab? With a brazen lack of irony, we came to believe that we could become a leader in our industry; a company that people hammered at our doors to work with or to work for.

Tibbitts Tip
Legendary Cultures

Our top competitive advantage was that we had created a company that people were beating at our doors to work for and with us. This was the special sauce of Starsys' success. Given the shortage of—and importance of—A-team talent to the organization, this became our company superpower.

This was in part because of our product excellence, but more so, our culture of fun, family, and honor of our team. It meant the very best came to work for us. They did not have to be "bought" into the company, and we had practically no attrition. Attracting that kind of talent brought ten times the value of the cost of making the company special. The ***morning meeting*** was at the nexus of the culture and the platform upon which much that was special about the company manifested.

FROM THE GARAGE TO MARS

Defining the traditional elements of a company's values, such as integrity, excellence, and customer service, can be a straightforward process of rewarding the behavior you are looking for and discouraging the opposite. What galvanizes company cultures from good to great are revered, legendary traditions that tend to cantilever out over the abyss of common practices. They are meant to bring companies right up to the bright-white line between what they are and what they want to be. By definition, they are not common practices or easy ways of being. Instead, they tend to be edgy and manifest a quiet thrumming of, "This is pretty far out-there. Should we really do this?" The oddest things become these elevated traditions. Not plotted or planned, but something done once, followed by "Maybe we should do that again."

Some of the more edgy cultural cornerstones replicate poorly outside of their birthplace without nuanced re-crafting a result of the unique chemistries at play within a founding team as it addresses tactical challenges through the lens of a shared, declared culture and personal values. The combination of cultural spices that make the special sauce of one company does not lend itself to a singular, easy-to-follow recipe of another. Transplanting a company-changing tradition of one company to another can fall flat like a shaken soufflé when exported "as is" without tweaking. Picturing GM executives zipping around headquarters on Segways as does Google leadership is like imagining Shaquille O'Neil as the lead in *Swan Lake*.

We began to realize that the processes and games we implemented (from here forward called "Games") were a shortcut to quick and powerful cultural change, further driving culture into our company DNA, as well as providing a crash course for new hires as to how we worked at Starsys.

The rocket launch celebrations and Vomit Comet flights were two traditions at Starsys that epitomized our cultural DNA of fun,

THE MORNING MEETING

> **Tibbitts Tip**
> **Game Design**
>
> Starsys' Game design strategy was modeled to a degree on a concept that Jim Collins, author of *From Good to Great*, posited "Catalytic Mechanisms," a process put in place that creates catalytic change simply by following the rules of the process or Game. By way of example, of the Games we designed one that was simple in concept, challenging in execution, and powerful in result—the **Profit Game**. Every month, we presented our profit numbers to the company, sharing profits with the employees of 10% of the profit we made in the month if the profit was between 0% and 10%, and 15% of the profit if above 15%. We had an open-book philosophy, and educated the company on what they did that created profit or loss. Setting it up (particularly making it clear how each contributed to profit) and how best to educate and present, was complex, but the result was powerful. We had 150 people aligned, doing everything they could to help the company be profitable.
> Another transformational Game was **Efforts and Blunders**, discussed later.

but the mother of all Starsys company culture was our *Morning Meeting*.

The initial idea was to hold a crisp, short, all-hands meeting once a day to maintain the week's action item list, a natural reaction to coordinating the work of eight people when we were a small company. The meeting developed its horsepower during the first year, when we began to use it to publicly declare program actions for the coming days. In essence, we were doing a form of Scrum before Agile project management had been invented. We purposefully encouraged a context of the program managers using the meeting to share their high-level to-do list for the program. They would, of course, include key milestones such as the delivery of hardware, but also, the "I've got to get with Phil to understand how he worked with this particular customer" tasks.

A whiteboard was placed in the front of the room with each program having its own area on the board. Actions were posted by the program managers, along with the responsible party's initials and the "committed-to" date. Anything could go up on the board, major or minor, as long as it had a date. Humor was encouraged. Peer pressure provided impetus for folks to get actions on the board; to not do so was to imply that work was not being done.

All of us knowing everything that was going on throughout the company forced a flat organizational hierarchy that had amazing torque. As we grew, we became unwilling to give up the organizational omniscience it provided. We committed to morphing the Morning Meeting into being relevant and effective for what was to be a 30-person company and then later, a 50-person and a 100+ person company. For it to work, the system needed 100% buy-in and a structure that was ridiculously efficient in organizing chaos and managing time. Many things that were tried eventually failed; but when we found something that did work, we stuck with it.

Here is a quick snapshot of some things we learned did not work, as well as some that did work:

Things that DO NOT work for getting a meeting to start promptly at 8:00 am.
(1) A heartfelt plea that timeliness is next to godliness.
(2) $1 per minute penalties for latecomers
(3) Playing the theme from *2001: A Space Odyssey* with the expectation that everyone is in the meeting room by the end of the music.

Things that DO work:
(1) A company-wide bell that rings prior to the meeting, and lasts exactly 60 seconds
(2) A "quarter to the party fund" penalty for all who aren't in the room by the time the bell ends.

We cycled through everyone in the company as meeting leader; each having the "opportunity" to run the board for a week. The clear message was that the board was by and for the team.

The board documented the 30 or so programs and the handful of actions that each required to be completed in the coming few days with the initials of the responsible parties noted for each task.

THE MORNING MEETING

There were as many as 100 actions on the board at any one time, organized by program. As the meeting leader got to the specific program on the board, the relevant program manager might be heard shouting out, "First action is complete, second is moved out two days, third is complete. Also, please add a fourth, 'get machined parts to Jason' for this Friday with my initials." And then a few seconds later, the leader would move on to the next program.

Each day, a tally was made of what had been completed and what dates were moved out. Dates noted in green were for on-track agreements and those in red were for renegotiated or missed actions. The pulse of the company was there to see, as everyone became aware of each other's activities. This was the vital "cross-strapping" necessary to ensure a high-performance team and was exemplified by comments as, "Tom, I've worked with that epoxy material before and had some problems. Let's talk after the meeting."

We found that starting the meeting on time was only half the battle. How to limit it to only 15 minutes? A second bell was added that rang for 60 seconds at the end of the meeting. Suddenly, the meeting leader was managing the meeting to finish before the bell. Group pressure for the meeting leader to finish the meeting in the time given was a powerful motivator. The board managers, driven by the bell, became masters of efficiency. With their friendly exhortations to keep us moving along, we found that we could tag in on more than 50 actions in less than 10 minutes.

As an acknowledgement for running the board, the company took time at the end of the week for everyone to write a small note of thanks to the meeting leaders, with each of us pointing out what we most appreciated about them during the course of the year. These notes were collected by HR and presented to the individuals who had run the board. The notes were heartfelt and scratched a deep, widely shared itch to be noticed and appreciated for their contribution to the team. Even the most stoic board managers eagerly awaited HR presenting the envelopes, each containing 75 personal notes sharing what was particularly appreciated about that individual. Like many other Starsys team members, both current and past, I still have each of those bundles of feedback that we called "rainy day notes" that I will occasionally open and reread. They continue to fill my cup in ways few other things do.

One of the key benefits of the board was that it provided a way to measure the day-to-day agreements made between coworkers, such as "Anne, I'll get that to you by Friday." As a result, we were now able to monitor the percent of the informal day-to-day agreements that were being kept between people. This gave us a window into the agreement integrity of the company, and we started posting the track record for all to see.

Year after year, the number hovered around 50 percent, and that just seemed low. If this number was too high, we knew it could mean we were losing our nimbleness. But only 50 percent? We were sure we could do better than that.

> **Tibbitts Tip**
> **Birthday Notes**
>
> When we were a smaller company, to celebrate birthdays, we would surprise the birthday person by popping into his or her cubicle with a cake, a song, and joke gifts. As we grew, this became unwieldy. It evolved into my writing a personal note to each employee on that special day. (Our HR director reminded me of whose birthday was when.) The note included particular gifts I saw in them, thanking them for working with us for the past year. This was a powerful way to convey that they were both noticed and appreciated

Note to self:

Things that DO NOT work for raising the agreement integrity of the company.

(1) Everyone in the room making a loud buzzer sound every time an agreement is missed.

(2) Daily tracking of agreements and posting of the results with a clear goal posted on the graph, and a daily focus on missing the mark once again.

THE MORNING MEETING

We tried everything we could think of to bring this number up. Nothing worked. Was this simply a law of organizational behavior that would prevent us from ever reaching our goal? We were not yet ready to throw in the towel. We decided to go way out of the box.

I stood up in front of the team during the meeting one morning. "I've got a deal for you," I said. "If our company can exceed 75 percent on the board and hold it for two weeks, we will bring in two masseuses on Friday. We will set them up in a conference room, and everyone in the company will get a massage. Thereafter, every two weeks that we are above 75 percent, we will do the same thing."

Much discussion followed—most of it the "You've got to be kidding!" kind. But it was too good a deal for the company to dismiss. I'm sure half of the motivation for those who questioned the offer was simply to call our bluff and see if we would keep our word. I was personally skeptical if the manipulation would work.

Such declarations are big bets. If they work, a company takes a giant step in raising the bar for performance, as well as elevating its culture in general, leaving a foundation for similarly big bets in the future. If they fail, the ability to declare bold organizational development experiments in the future is handicapped by redoubled skepticism. You only get a couple whiffs in swinging for the fences before experiments like these are dismissed out of hand.

It took 24 hours for disruptive change to occur. The next day, the company went from 50 percent kept agreements to 77 percent. Folks were on the edge of their chairs as we worked the board each day and at the end of the two weeks, the metric was solidly above 75 percent. We looked for sandbagging, but it wasn't there. The quality of the actions had not changed. Individuals or individual teams were keeping their agreements.

Good to our word, we set up the two masseuses and everyone got a ten-minute massage.

FROM THE GARAGE TO MARS

Was it disruptive to schedule the massages? To an extent. Was it expensive? Yes—both the cost of the masseuses and the lost time added up. Was it worth it? Absolutely. The value in increased efficiency more than paid the bill. It also created a morale boost by demonstrating appreciation for exemplary performance.

There was also an ancillary benefit that was subtle yet important as well: Providing free massages was a prized, egalitarian reward available to all at Starsys, independent of salary or position. In short, it was an action that helped inoculate the company against elitism and hierarchy. It was a message opposite that sent by such dated practices as reserving the best parking spots for senior execs or corner offices only for the boss.

The massages continued after the first week's experiment, as we hit the mark about three of four times. To keep costs low, we used masseuses from the Boulder Massage School across the street. We initially offered massages only for our employees, but after a few months, we began offering them to visiting customers as well, creating an intense competition among them to be the company representative chosen to attend Starsys product design reviews. This action ended up feeding our growing reputation in the industry. We were the company you wanted to work with or for.

Some traditions have staying power, maintaining their significance for decades, but the morning massages were not that. They were a bright flame that flared for a short time, served their purpose, and faded away. After a year and a half, a sense of entitlement and familiarity settled in. It was time for a change, and the massages ended. But the 75 percent agreement result had become a permanent part of our culture. A significant shift had happened and a new habit created that lasted for years.

Over time, the Morning Meeting became a locus of company culture in many ways. Through it, new traditions were born,

THE MORNING MEETING

company-changing discussions were held, and critical values debated. Other key parts of our culture were birthed at the meeting, and over time, these became bold statements of how we worked together. You needed to do nothing more than attend a couple of Morning Meetings, and you knew that Starsys was unlike any other space company out there. Here are just some of the elements that became legendary inside and outside of our four walls.

CODIFYING VALUES

Ever since the book *In Search of Excellence* by Tom Peters came out, company values have been in the spotlight. Choose and declare your company's norms and ethics as the guiding principles for the company.

This is spot-on in concept, but often poorly executed. A poster on the wall that states "Integrity is the key to our success" or "Innovation is what makes us great." Not a bad thing, but by itself has little impact on company culture.

This is much less of an issue when a company is small, as the culture is organically ingrained and practiced; it just "is." But as the company grows, it is critical these be memorialized and codified, particularly since a value such as "integrity" can have different definition to those who started with the company than the new hires.

I decided that we should try a grand experiment with the codification of our values, and stumbled onto an approach that was exceptionally powerful. The approach was to balance of two approaches in favor at the time: a) Management defines and declares the company's values from on high, and b) the employees are tasked with codifying the values they believe the company stands for.

I met with the leadership team and came up with the names of six values we believed were key to our success:

1) Excellence
2) Innovation
3) Fun
4) Family
5) Integrity, and
6) *Otlechno*, a Russian word for finding the perfect balance between conflicting needs.

We then asked the company to define each item. We dedicated a weekly one-hour meeting to the process. As a group, we took on one of the values each week, and as a group brainstormed and bulleted the definition for each, encouraging healthy discourse. We strove for definitions that were unambiguous and not only described what the value was, but what it was not. For instance, Fun was defined in part by "It is making work fun" and "It is NOT fun at others' expense."

And then the company voted on every word of every bulleted definition, with us iterating until we had unanimous agreement with every agreement. A single objection to a bullet led to additional discussion until there was complete agreement and alignment.

We then had the values written on a large framed poster, with everyone in the company signing the back; it was posted prominently in the front office. In addition, we had wallet-sized cards printed up with the definitions. We passed those out and encouraged the employees to keep in their wallets or purses, which most did.

The company *owned* those values, and they were used as the basis of operations and decisions. When we began to get into the defense business, one of the team stopped me in the hall and said that he didn't think it was the right thing for the company to be involved in defense, that it was against what we stood for. I gave his comment consideration and suggested we check. We both pulled out our values statements, and found that none conflicted with working in the military. He said, "Yup, it is aligned with our values," dropped the issue and never brought it up again.

I was reminded of the power of the values ten years later after Starsys had been sold. I ran into Terry in a brew pub. Terry was a technician I hadn't seen in those ten years. We started reminiscing, and at the end, he said "I've got to show you something" and pulled his wallet out. From that, he drew a wrinkled, laminated, 15-year-old card. It was the values card we had handed out way back when. "Scott, I still have this as a reminder of how a company can be."

At the same time, looking back, the process was more onerous that it needed to be. We had taken the process of creating and owning values to the extreme, and I was proud of how it had

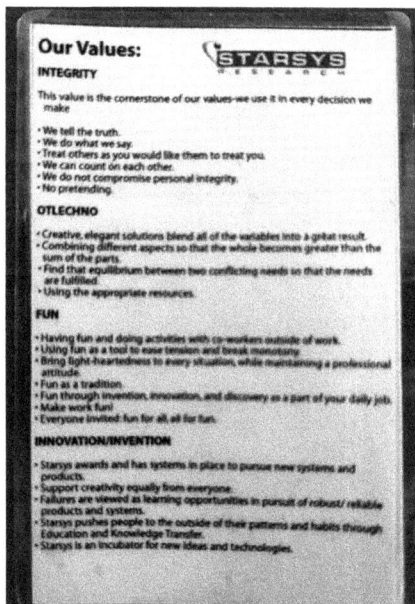
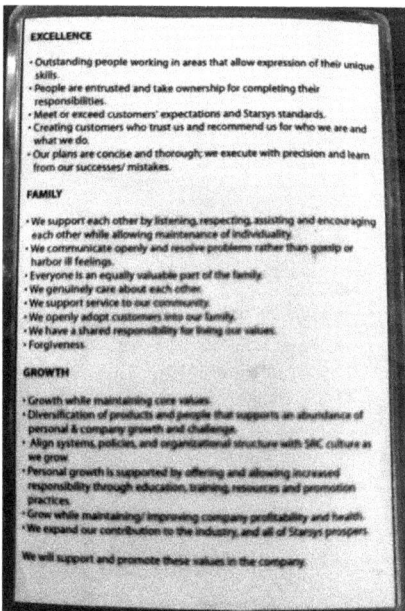

STARSYS' VALUES

INTEGRITY
This value is the cornerstone of our values—we use it in every decision we make.

- We tell the truth.
- We do what we say.
- Treat others as you would like them to treat you.
- We can count on each other.
- We do not compromise personal integrity.
- No pretending.

OTLECHNO
- Creative, elegant solutions blend all of the variables into a great result.
- Combining different aspects so that the whole becomes greater than the sum of the parts.
- Find that equilibrium between two conflicting needs so that the needs are fulfilled.
- Using the appropriate resources.

FUN
- Having fun and doing activities with co-workers outside of work.
- Using fun as a tool to ease tension and break monotony.
- Bring light-heartedness to every situation while maintaining a professional attitude.
- Fun as a tradition.
- Fun through invention, innovation, and discovery as a part of your daily job.
- Make work fun!
- Everyone invited fun for all, all for fun.

INNOVATION / INVENTION
- Starsys awards and has systems in place to pursue new systems and products.
- Support creativity equally from everyone.
- Failures are viewed as learning opportunities in pursuit of robust/reliable products and systems.
- Starsys pushes people to the outside of their patterns and habits through education and knowledge transfer.
- Starsys is an incubator for new ideas and technologies.

EXCELLENCE
- Outstanding people working in areas that allow expression of their unique skills.
- People are entrusted and take ownership for completing their responsibilities.
- Meet or exceed customers' expectations and Starsys standards.
- Creating customers who trust us and recommend us for who we are and what we do.
- Our plans are concise and thorough; we execute with precision and learn from our successes/mistakes.

FAMILY
- We support each other by listening, respecting, assisting, and encouraging one another while allowing maintenance of individuality.
- We communicate openly and resolve problems rather than gossip or harbor ill feelings.
- Everyone is an equally valuable part of the family.
- We genuinely care about each other.
- We support service to our community.
- We openly adopt customers into our family.
- We have a shared responsibility for living our values.
- Forgiveness.

GROWTH
- Growth while maintaining core values.
- Diversification of products and people who support an abundance of personal and company growth and challenge.
- Align systems, policies, and organizational structure with Starsys Research culture as we grow.
- Personal growth is supported by offering and allowing increased responsibility through education, training, resources, and promotion practices.
- Grow while maintaining/improving company profitability and health.
- We expand our contribution to the industry and all of Starsys prospers.

We will support and promote these values in the company.

worked. Reflecting, we could have created as effective a result with a streamlined version of the process.

THE GRIPE BOX

The box was built to be gritty. The look was intended to encourage submitting problems within the company that were getting under people's skin. This wasn't your father's politically correct suggestion box—the Gripe Box was designed to solicit "...and another thing that pisses me off ... " feedback that allowed someone to discharge built-up angst and emotional energy. It also enabled the company's leadership to quickly find the proverbial burrs lodged under employees' saddles.

Once every month or so, Jenny, our HR director, and I would steel ourselves, open the box, and read the anonymous notes within. The challenge of any suggestion box is getting folks to contribute, and the key to that is being responsive to the contributions. The

The Gripe Box

FROM THE GARAGE TO MARS

Gripe Box's success was a result of disciplined follow-through. Wimpy or non-existent responses from management quickly create dusty and unused suggestion boxes with a few gum wrappers in the bottom for emphasis. Such was not the case with our Gripe Box. By committing to creating substantive, real-time responses to each gripe, we created a genuinely supportive climate for addressing concerns of "be thoughtful of what you gripe about, as it will be addressed," as opposed to a hostile vibe of "Why bother? Nothing ever changes here."

At the end of each month, we read a dozen or so notes, ranging from topics easily addressed, such as "The bathrooms are too cold," to challenging issues needing attention to inappropriate vitriol that was gently diffused. We found that what made the Gripe Box succeed was transparency, sincerity, and action.

Prior to the meeting, Jenny and I would read each note, strategizing on how to best address each issue. At times, the vitriol in the notes made them a bitter pill to swallow. An example:

> "How much longer will we keep trying to maintain the illusion that our salary review process is honest. We claim that raises are based on merit ... I cry bullshit on that!"

The success of the Gripe Box was based on our willingness to share this feedback in its unvarnished form with the company. We did this by my reopening the box in front of the team during the meeting and reading each note. Now and then, there were comments that were not appropriate to read publicly, such as when someone complained about a particular person, and in these cases, we dealt with these privately. But 95 percent of the notes were read as is and addressed extemporaneously in front of the room.

Oddly, the transparency of the process created a discomfort that worked in our favor. Being in the audience when we opened and read the Gripe Box notes was a bit like going to a demolition derby.

THE MORNING MEETING

There was both a cringing and a morbid curiosity to witness the "The Emperor Has No Clothes" theater that occasionally arose a from the CEO reading notes that exposed leadership flaws.

As long as the notes didn't involve a personal attack, we read each one, bringing as much real-time accountability as possible to the discussion. The importance of this type of leadership had earlier been affirmed by someone I am fortunate to know as a friend, Art Stephenson.

Art was a director of the Marshall Space Flight Center (MSFC) at the time, and he shared with me a key piece of leadership wisdom gleaned shortly after he took on his NASA Director role. Previous MSFC leadership had made a blunder with a new initiative just before he took on his new role that negatively affected the Center's employees. The result was a morass of bad feelings that significantly impacted MSFC morale.

As it turned out, Art chose to publicly apologize to thousands of MSFC employees about the screw-up that had happened to this organization now under his watch, even though he had nothing to do with it. In telling this story, he surprised me by saying that this was the best possible scenario that could have happened to him as a neophyte director. He was able to galvanize the troops as a leader by showing what he stood for in a tough situation. By doing this hard thing, apologizing to thousands, and stating that he would make good on the blunder, he showed unmistakably that he walked the talk of the values he was promoting by doing the right thing by the Center's employees.

Similarly, each reading of the Gripe Box gave us a slew of challenges to address as the company watched carefully how we would do our version of the Right Thing.

There is a principle of the Japanese martial art of Aikido that captures the philosophy that we were inadvertently manifesting:

FROM THE GARAGE TO MARS

The foundation of the self-defense aspect of aikido is the act of redirecting the attacker's energy, rendering it harmless or even beneficial. Turn the negative situation into a positive one, a curse into a blessing.

By soliciting gripes from our people, we were welcoming an attack of sorts from which we would strive to nimbly pivot, innovating in real-time to flip the negative into a positive, blocking and changing the sentiment of a gripe into a force for good, without a loss of momentum.

A good example of this was the morning we received an angry complaint about the handful of smokers in the company. They were taking an extra 15-minute break a couple times a day when they headed out for their smoke, not to mention disrupting the work of their teams. As I read the note that claimed this was unfair, I could see dozens of heads nodding in agreement, reflecting the frustration with this situation.

After reading the note, I said that we expected anyone who took a break for smoking, or who took time off for any other personal needs, to make up this time.

I then walked to the whiteboard to introduce an idea that was the result of a flash of inspiration that Jenny and I had when we first read the note. I wrote "Starsys $500 Challenge" on the board and explained that any smokers in the company could put their name on the board to declare that they were quitting smoking, committing to not smoke for at least six months. For any smokers who succeeded, the company would pay each one $500. Several hands in the room went up immediately. Over the next couple of days, more people put their names on the board and more names were added over the next few months. Eventually, more than ten people were participating in the challenge, and we ended up writing checks to more than a half-dozen people whom we helped quit smoking.

THE MORNING MEETING

The $500 challenge was a win-win financially, as the reduced number of breaks taken more than offset the rewards that were paid out. Of greater importance, though, was the fact that this incentivized challenge created empathy for smokers in the rest of the company, helping their coworkers see that the breaks were more about a battle with addiction than a perk. Similarly, the smokers realized the impact of their smoking on the company, and they recognized the willingness of the company's leadership and their coworkers to help.

What created the most impact, though, was Starsys' motivation at the heart of the solution. More than productivity, we really did want our team to be healthy and have lives that worked, and we were willing to invest hard dollars toward that result. The solution diffused the energy in directly addressing the heart of the problem, and it allowed us to put our money where our mouth was when we said, "Starsys is a family that takes care of each other."

Did everyone who got a $500 check stop smoking for good? Nope. About a fourth of those who got the reward later picked up the habit again. And about a fourth of the group started smoking again during the initial six-month period and didn't get a reward. But there were about a half-dozen Starsys employees who quit for good with longer, healthier lives as a result.

THE POWER OF PAYING IT FORWARD

When leadership commits to honoring individuals in the company, the individuals in the company find themselves looking for ways to return the honor. The power of this "honor those around you and they will honor you" ethos cannot be underestimated in a company. Every employee in a company has unique gifts, and when they look to manifest those gifts to give back to the company, the results are spontaneous, unexpected, and powerful.

FROM THE GARAGE TO MARS

An early example of this occurred in the early '90s when the Internet was coming to life. At the time, it was a novelty that was becoming a hobby for some of our technicians, but it was something not quite yet necessary for business. As time passed, we were oblivious to the increasing importance of having both a website and a killer domain name. I realized in the spring of 1992 that we might have missed the boat, and I rushed to task someone to grab the domain name starsys.com. My fears were justified, as the report came back that the domain name had been taken more than a year earlier. I settled into the idea that we were going to have to pay handsomely for rights on the one domain name that made sense, and I asked that we track down the owner and open a dialogue.

The response came back that this could happen quickly, as the registrant was a Mr. M. Richardson, who lived in a nearby town. A thought crossed my mind ... we had a technician named Mark Richardson. Could it be? I walked back to his desk and asked him if he knew anything about starsys.com. His response was unexpected. "Oh yeah, that. A couple years back, I realized we would need the domain name but figured you wouldn't understand why. I just went ahead and got it registered for us. Here's the account info," he said.

As with Kyle Hickey carefully locating the Starsys logo to be seen on the first picture back from Mars (see the *On Mars* chapter), Mark had done something similar, creating a surprise gift for the company, one that took multiple years to be understood and appreciated.

EFFORTS AND BLUNDERS

When we hired people, a part of the ingress process was their meeting with me, the CEO, for a one-on-one so that I could share the key values of this unique company they had just signed on with.

---

[28] The attentive reader will notice a discrepancy between the acronym and the politically correct definition.

THE MORNING MEETING

This included my conveying a key tenet of the company, best expressed with the acronym:

IANSHTWIS

"It Absolutely, No Kidding, Has to Work In Space"[28]

If I were to articulate this mandate a bit more expansively, the new hire would likely have heard something like the following:

"You need to be willing to do whatever it takes to ensure our stuff works as promised, as every one of us is responsible for our hardware working in space. If it ever does not, a billion-dollar mission could be lost, or an astronaut killed, and our company would be out of business."

Nevertheless, now and then, technicians or engineers would make a major gaffe, and they would find themselves wondering if they would be let go. When I heard of their mistake, I sought them out, not to admonish, but to ease their angst by empathetically sharing the story of my first major mistake at Starsys—when I put actuators in a too-hot oven, ruining them the day before they were to be shipped.

The purpose was two-fold: to let them know mistakes can happen as a part of doing something as challenging as rocket science, as well as to convey that framed properly, mistakes can have a silver lining when used discover lessons learned, lessening the chance of the same mistake being repeated by others in the company.

We manifested this idea of the instructive and redemptive value of mistakes on a grand scale with "Best Efforts and Blunders," an

---

[29] A long, thin, flexible tube with a lens at one end and an eyepiece at the other that can be threaded into small enclosures to allow inspection of places you otherwise wouldn't be able to see. You knew the technicians were in a playful mood when you heard the occasional, "Anyone else want to take a look at my tonsils?" comment float up out of the inspection room area.

event held in an extended morning meeting once a month. Anyone in the company could nominate the most notable heroic efforts they had witnessed over the previous month, such as "Kevin worked through the weekend and made sure that the documentation was completed allowing us to ship Monday." The company would then vote on the Best Effort from the nominees, and winners were awarded a gift certificate and had their names put on the wall.

This was standard stuff for any company that honors heroic efforts. However, we added a heretical twist: We created an equivalent award for "Best Blunder." With sarcastic humor reminiscent of a comedy roast, coworkers would nominate each other for the biggest gaffe made during the past month. At times, these were significant mistakes and costly to the company. The bigger the impact to the company, or the more harebrained the mistake, the more likely the nominee was to win the award.

One winner included an individual who dropped our only borescope, requiring us to buy a $2,000 replacement.[29] Another example was the technician who improperly cured epoxy, costing us thousands of dollars when a group of parts had to be scrapped. However, there was method to our madness: by celebrating these blunders, we were emphasizing how important it was in our business to not hide mistakes. If someone dropped a piece of spaceflight hardware on the floor when no one was around to see, we had to have that person be comfortable reporting the event. More importantly, it was vital that lessons learned were passed on to the rest of the company to prevent a similar error from occurring again. The underpinning of this thinking was expressed as:

> *"We are all talented individuals who care about what we do. But being human, we at times will make mistakes. When these happen, we will correct the error and share what we learned with the rest of the company."*

THE MORNING MEETING

Unspoken, but nevertheless understood, was that a pattern of an individual winning the Biggest Blunder award more than once would not bode well for that person's future in the company.

By not only celebrating but also by actually creating a competition for the biggest blunder, we unmistakably drove into the bedrock of our culture the understanding that mistakes will happen when you shoot for the stars. However, the important thing was to correct mistakes quickly with integrity and not to make the same ones twice. To this end, part of the deal for winning the monthly "Biggest Blunder" was sharing with the company what was learned from each mistake.

To further underscore our commitment to recognizing notable efforts as well as instructive mistakes, we created a tradition at the yearly Christmas party in which we wrote each blunder and effort on a small sheet of paper, put them in a hat, and had one blunder and one effort chosen at random to create two Grand Prize Winners. The winner was announced, along with a brief synopsis of the particular effort or blunder and awarded a gift certificate for the employee and spouse for either a hot-air balloon ride or a weekend getaway at a bed-and-breakfast. This final element of our Best Efforts and Biggest Blunders, rewarding the family rather than just the employee, was a powerful way to emphasize another core part of our Starsys DNA: That the family at home, supporting the employee, was also considered an important part of the Starsys family.

Fortunately, a clear measure of success for our investment in recognizing notable efforts and blunders was readily available: the success rate of Starsys hardware in space. At the end of the day, each piece of hardware was dependent on hundreds of intricate steps that had to be executed flawlessly for our devices to work as promised.

Starsys cumulative success rate over 25 years was unprecedented and became legendary: more than 3,500 devices flown in space with *zero* failures.

The result was deliciously ironic. By celebrating our mistakes and the lessons learned, we ensured that every device that went into space with our name on it was flawless.

BEST TEAM BUILDING OFF-SITE ... EVER

Mention "team building off-sites" to virtually any audience of adults and you can pretty much bank on the reaction. Eye rolling for sure. Maybe a few snarky comments citing a *Dilbert* strip or an episode of *The Office*, followed by bursts of derisive laughter. Or even a few bars of "Kumbaya" offered up in hushed ridicule.

We tried pretty much every team-building exercise out there. These included trust falls, as well as imaginary alligator-infested rivers crossed by too many team members balancing on 2x6 planks and small plywood squares. Some of these were moderately successful, others less so.[30] The Gripe Box did a pretty good job of letting us know what the collective mindset was. While there were comments that the time away in the mountains was appreciated, the results overall were best summarized as ... lame. By the time we had grown to a 50+ person company, we decided it was time for a change.

Instead of simulating situations that would create little more than a caricature of how we worked together, we came up with something that we thought would force true teamwork: Spend the off-site budget on an investment in a day-off, day-long project for the community.

---

[30] Lesson learned. Do not use paintball as a team-building event unless you are looking to create a vengeful win-lose culture. Technicians shot at close range in the thigh by an engineer will require quite a bit of coaxing before they will once again work on weekends to help said engineer out

THE MORNING MEETING

Being a fun, mechanically oriented company, the idea of building a playground for a school in need seemed like a perfect fit. And as it turned out, finding the right project ended up being surprisingly easy. After a couple of initial dead ends, we called the Social Services Department in Boulder, which quickly found a recipient for our community-minded goal. There was a daycare center in a nearby town that supported lower-income families and was desperate for playground equipment but short on improvement funds. After creating an off-site committee to get things up and running, we made a visit to interview the daycare center and declared it a great fit.

A couple more visits were made by the off-site committee to line up what needed to be done and what materials would be required for the project. In addition to purchasing and installing playground equipment, the tasks included landscape work, furniture assembly, painting and computer set-up. Starsys contributed the money for the playground hardware and related construction materials and also provided the labor.

Starsys Research Team and the playground equipment built in a day for the Boulder Women's Shelter.

FROM THE GARAGE TO MARS

Tibbitts Tip
Make a Difference—Off-site Design

- Pick an activity that matches what the company or group does for work. In our case, assembling things turned out to be perfect for a company that builds space mechanisms.

- It is not at all difficult to find a worthy cause. With a few phone calls, you should be able to find someone at a social services agency who is aware of organizations or people in need. Once you find these people, the momentum builds.

- Interview candidate organizations and don't settle for anything other than a great fit. It is important that the work is badly needed, that your company can do the work well, and that the organization will be grateful.

- Have a wide variety of useful things to do, some of which don't require much direction. As people complete tasks, they will want to have something else to do. Also, make sure that some of the work is not too physically demanding.

- Don't fully organize the group. Value comes from the team self-organizing. It is helpful, however, to select a lead for most of the tasks (preferably people who are not normally in a leadership position in the company to mix things up a bit).

- Remember those old-style monkey bars we all used to swing on? They are cemented in holes that reach down to China. Take our advice. *Do Not* try to dig them out if you happen to take on a playground project. Lop them off with a Sawzall (this was our HR director's initial suggestion, which was ignored by six engineers with shovels for close to an hour).

- Cap it off with a celebration nearby to close out the day.

On the day of the off-site, we were surprised by how the daycare center really rolled out the red carpet, greeting us with tables of refreshments and homemade food prepared by the parents, letting us know that our help was truly appreciated. This set a great atmosphere for the day, and there was a perceptible shift in our collective mindset as we realized the impact that we were about to have.

We were initially concerned that putting 50 people on a project without much organization would be chaotic and unproductive. However, this turned out to be unwarranted. The group quickly

THE MORNING MEETING

self-organized into work teams and split up to take on about ten tasks simultaneously. Also, there were enough tasks to do that each of us could complete one job and then quickly move on to the next one.

One especially great thing that occurred was that a different work structure spontaneously developed with people who were skilled in particular tasks (such as patio laying or carpentry) taking the lead while helpers organized around them. This structure was independent of the company work structure so that in many cases company leadership was directed by those who normally worked for them. As a result, our people had a chance to demonstrate effective leadership in ways that normally didn't happen in their regular jobs, and hierarchy was turned on its head.

An astonishing amount of work happened that day. We were surprised to find that we had completed pretty much everything on our list, including substantial tasks such as laying pavers for a patio, building shelves into an empty closet, programming computers, and assembling two sets of playground equipment. When we wrapped things up at the end of the day, the kids came to try out the equipment and we got some great group photos of a bunch of tired employees with some very excited kids.

We capped off the day with a dinner at a local brew pub. My fondest memory is of a group of worn-out folks, swapping stories of the day over pizza and beer and not wanting the evening to end.

*****

A week after the off-site, a package arrived. Inside were handwritten thank-you notes from each of the kids and a set of pictures from the day. The letter from the director was read to the company and brought tears to many eyes. For weeks afterward, people kept talking about the "best off-site ever." For some folks in the company, it was their first taste of assisting a community in a volunteer role. From a team-building aspect, it was a slam-dunk win.

FROM THE GARAGE TO MARS

After that first experiment, there was no going back to trust falls. It became a part of our culture that once a year we would pick a cause, stop work, and help. One year, we found another pre-school Another year, we transformed a women's shelter. After the daycare project, our employees would hunt each year for the next great volunteer opportunity.

Our community service efforts were expensive. On paper, you could argue it reduced our annual profit by one-half of a percentage point. Was it money well spent? Absolutely. I believe that the one-half percent came back many times over in retention, company pride, teamwork, and the intangible return of giving back to the community.

MANAGING FOR THE FAMILY

In 2010, the Bureau of Labor Statistics reported that on average, an employee changes jobs 11 times between the ages of 18 and 42. That translates to an average tenure with a given employer of less than 2.5 years.

In the aerospace industry, as well as in other high technology, service-focused industries, a primary indicator of market-dominating success is the prevalence of employee talent with 5 to 20+ years of

---

Tibbitts' Tip
The Power of 3-Day Weekends

One of the most expensive Games in the company was alternate 3-day weekends. This meant 5 day/4 day schedule where all were expected to work more than 80 hours, but to take off every other Friday. We monitored the productivity of the company, and while we were getting more than 80 hours from people in the two weeks, with the extra day away, hours worked lessened.

Having a workplace where you get a Memorial-Day-type holiday every other week was life-changing. A day for each of us to get caught up on our personal lives, or to take a long weekend with family, or to come in to work for a few hours to get caught up. If you are looking for a way to ensure low attrition, this rocks that.

THE MORNING MEETING

experience in the company. These high performers who grow up with your company gain wisdom from battle scars and lessons learned.

Here is a maxim of sorts that can frighten but also excite: be great at retaining great talent and industry-leading success will follow. There is much to be said for a company strategy simplified to *Do One Thing Well* and to have it be a pretty good *One Thing*. There are also certainly many other strategies that have been tried to achieve retention success, such as incentive packages, stock options, and any number of generous benefits. Some are expensive and some work better than others.

Our strategy for retention was simply to strive to maximize the quality of life for the Starsys team. We worked to create the kind of company we wanted to work for. In the course of developing this guiding principle, we discovered a powerful strategy that tends to fly under the radar:

If you want talent to stay, manage for the family.

This evolved over the years to managing the company to be not only the company employees wanted to work for, but to be the company that the family was glad Mom or Dad worked for. It permeated our decision-making process—balancing the conflicting needs of financial return with the quality of life for our employees and their families.

Dozens of policies and practices came from this balancing act through the years. None by itself was earthshaking, but the sum total said much about what was important to us:

- Children and spouses were welcomed at our company. We took the time to get to know our families and explain our work, and we made it clear that a visit was welcome, not an imposition.
- We encouraged employees to take company time to go into their children's classrooms, whether it was to talk about Mars, perform Mr. Wizard experiments with liquid nitrogen, or tell stories.

FROM THE GARAGE TO MARS

- We worked to minimize overnight travel or the length of time that employees were away from their families. When possible, we were creative with travel budgets to support employees taking spouses or the family along to special locations.
- Recognition for performance honored the spouse as well: weekends away at a bed and breakfast or hot air balloon rides for two were typically awarded.
- We supported local schools by opening the doors for tours given by company leadership.
- We experimented with and then adopted a work schedule with every other Friday off, giving employees and their families a three-day weekend 26 times a year.

The result? For the first 15 years of Starsys' growth, annual employee retention hovered between 90 and 95 percent. At our 15-year anniversary, we counted 65 employees with five or more years of tenure. More than 20 of these had ten years or more experience at our company.

* * * * *

By any standard, the Starsys Christmas party was epic. A talent show with a manager on a unicycle. Great food. A rowdy band. A gag gift exchange that included a can of pickled herring so beyond its spoil date it had inflated and had to be kept away from sharp objects.

The evening was a keeper, one we knew we would all remember. After dinner, I stood in the warm glow of a great party to toast and thank the 150 employees and spouses who were there. When I got to the point of sharing the profit numbers for the year, I was astonished by spontaneous cheering. Not the "Since Jim the suck-up is clapping, I guess we all should" variety. The room just erupted with "Wahooo!" I was caught unaware by not only the enthusiasm but by my emotional response to it.

Toward the end of the evening, Mary went out of her way to track me down. She was the wife of a top project manager we had brought into the company a couple years back, and she motioned me to a quiet part of the room. "I just wanted you to know how much it means to us that Charley is working here." I brushed off the compliment with a quick "We're glad to have him..." She stopped me short and put her hand on my forearm. "No. You need to know that it has changed our family. Thank you. We are here for good."

As Christmas parties go, that evening was a high-water mark for the company. Looking back, there were many memorable moments, including the public celebration of the money made by our company. However, to this day what continues to stand out the most for me was Mary's private "thank you."

Later that year, Charley navigated us through a minefield of a program that saved us more than half a million dollars. Far more than the total of the cost of our family-oriented benefits.

Although not our original intent, through the years, we discovered that driving "quality of life" deep into our corporate DNA was also very good business.

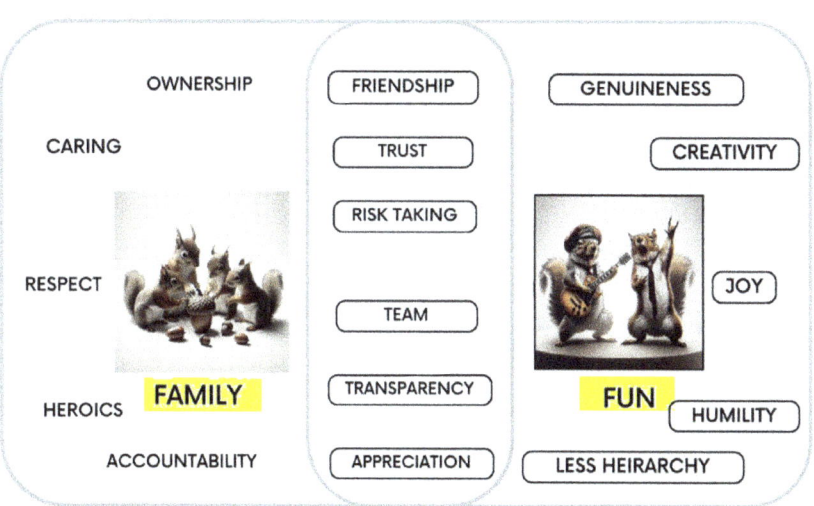

The impact of "F" words on a company's DNA

GRAFFITI

THERE MUST BE A DEEP primal need that drives us to carve our names on trees or scratch initials on mountain tops, as if in doing so we are declaring the range of our influence in the world. It is the yin to the yang of taking a picture to capture the where, conversely, graffiti is a way of showing to anyone who might happen to come along, or, in a larger sense, to the greater universe that "I was here."

Following that cold call I made to NASA in 1987, the agency invited me to meet with them because of a flagship mission that eventually came to be called Cassini, named after John Dominique Cassini, the French astronomer who discovered four of Saturn's inner moons. Cassini was a multi-billion dollar program with the objective of sending a nuclear-powered spacecraft to orbit Saturn for a decade to take pictures and perform various planetary science tasks. The ultimate goal was to better understand a planet that we only knew from fuzzy telescope pictures and a couple hundred images taken as Voyager 1 and 2 streaked by it in the early 1980s.

Around the time of my cold call, NASA JPL had desperately wanted to replace the explosives it had used on previous spacecraft with something more benign to open instrument covers. So, when I contacted the agency to recommend our wax actuators as a solution for its problem, this turned out to be a stroke of good fortune for both NASA and our fledgling company. By the time Cassini was built, Starsys had designed dozens of mechanisms and actuators for spacecraft, mostly for opening covers on instruments, including aperture cover mechanisms that opened instrument "lens caps."

FROM THE GARAGE TO MARS

We at Starsys were all keenly aware of what was at stake with this mission. If Cassini succeeded, hardware that we had carefully crafted in our little spacecraft company in Boulder would be orbiting a planet 934 million miles away.

Imagine the siren call we felt: a bunch of nascent rocket scientists thoroughly captivated by making stuff to be sent to deep space, having an opportunity to put our mark not on a sandstone wall in a national park but on a distant planet. Sending Starsys graffiti to Saturn was simply too compelling an opportunity to pass up.

But, which to choose? We had a half-dozen devices flying on the spacecraft, but we needed an accommodating customer that wouldn't dis our desire to put our mark on the universe. We chose an instrument cover mechanism that we were fabricating for one of our customers, Southwest Research Institute, and that had enough surface area to accommodate the initials of everyone in the company. And so, with permission from this customer to personalize our mechanism, and with much ceremony prior to beginning assembly, we set the covers out in the clean room and began vibra-etching our initials into the metal. Those of us who were parents included the initials of our children, setting the stage for the best–show-and-tell-ever opportunity for them down the road.

In a stroke of bartering genius, we invited the director of sales of a nationally known golf equipment provider, whose warehouse happened to be across the street, to see our Saturnian hardware. As it turned out, Ed's visit sealed quite a deal for our employees. In exchange for including him and his family on the cover, he agreed to provide everyone in the company with golf equipment at 80 percent off retail. Dozens jumped at the chance, and a week later,

GRAFFITI

our meeting room was quickly transformed into a warehouse with box upon box of golf equipment stacked against the walls.[31]

In many ways, the Saturnian graffiti and the golf deal were small things. They required little effort once the ideas were formed, but the impact of these statements reaffirmed the message that we continually drove home: Starsys cared about our team-family and was constantly looking for ways to give back. We had little idea how strongly that investment would pay off in the decade to come.

With a mission of this magnitude, the building and launching of Cassini and the travel time to Saturn made for a long journey. Cassini had been the impetus for the creation of our company in 1987, but the hardware that we provided was not actually designed and shipped until 1994. Cassini was launched in 1997 and then it spent seven years making the trek to Saturn. From inception to its eventual arrival at the planet, this spacecraft that was so instrumental in the creation of Starsys did not reach its destination until 16 years after our company was born.

The long journey was partly so because Cassini at first glance appeared to start by going the wrong way.

The spacecraft was actually launched *toward* the sun, performing two fly-bys of Venus to pick up speed. It followed this with a swing past Earth and then Jupiter to pick up even more speed before eventually arriving at Saturn in 2004. This "long way around the block" was necessitated because getting to a planet at this distance is all about conserving fuel. You have to get your spacecraft going crazy fast to get it all the way to Saturn, and then you need to burn a lot of fuel to slow it down once it's there to make sure it doesn't just

---

[31] Shortly thereafter his company filed for Chapter 11 (unrelated to our barter deal), and Ed moved on to another city and then overseas. The plaque that we created for him with pictures of the Cassini launch and his initials on the instrument cover sat in my drawer for 10 years. It gnawed at me for a long time that we hadn't been able to complete our side of the bargain, but I was finally able to track him down via Linked-in. I chuckled, imagining his response to the subject line of the email that I sent: "I've got something for you from Saturn." We met in Boulder a couple of months later and I was finally able to complete the deal that we had made years earlier.

zip by and head out to the boonies of interstellar space. Although fuel isn't that expensive by itself, it costs a fortune to get it up into space, with launch costs on the order of $10,000 per pound. To put this in perspective, it's like paying $2.50 for a gallon of gas but having to spend $80,000 to ship it to where you need it. . .

The "cheat" NASA engineers use to end-run this logic is called a "planetary fly-by." This is a technique to save fuel by stealing speed from planets.

Cassini's circuitous route had been meticulously planned so as to swing by as many planets as it could on its way to its final destination. The idea was for the spacecraft to swing in low around a planet and then to be hurled into space as if flung from a slingshot, gaining thousands of miles an hour in velocity while slowing the host planet's speed by an immeasurably small amount. Cassini had so far to go and needed to go so fast to get there that it required swinging by four planets to gain the necessary speed, ultimately accelerating the spacecraft to 53,000 mph. However, even at that rate, it took seven years to complete the journey.

On July 1, 2004, Cassini arrived at Saturn, firing its single engine for a nail-biting 96 minutes in slowing into orbit around the planet. Just one hiccup of the engine during that time would have caused the spacecraft to sail past Saturn and a $3.3-billion mission would have been a failure. But the engine worked flawlessly during those 96 minutes and Cassini (and dozens of our initials) made it there safely.

The discoveries that Cassini has made in the 15 years it orbited Saturn are mind-blowing, and possibly the most stunning example is Enceladus. Backlit images taken of this small moon (not much larger in diameter than the British Isles are in width) show thousand-mile-high geysers of warm water bursting from the tiny orb at hundreds of miles an hour. More than 300 pounds of water were being ejected into space every second. Think Old Faithful on

GRAFFITI

Enceladus and its thousand-mile-high, warm-water geysers. (Courtesy of NASA)

an epic scale. Turns out that the tidal forces on this moon are so strong that a warm-water ocean has formed under a thick outer surface of ice, resulting in ice cracking at the moon's south pole from these forces with the spray turning into ice crystals that form the planet's outermost ring. Most significantly, this warm-water ocean may have all the stuff in it that life needs to thrive. Because of Cassini, Enceladus is now considered the most likely place in the solar system outside of Earth to harbor life. NASA is looking for ways to get another spacecraft there to learn about it.

Cassini also took what is considered by many to be the most profoundly beautiful picture of our solar system. Timed with extraordinary precision, it is a picture of a Saturnian eclipse, the planet and its rings backlit by a hidden sun. Beautiful beyond description in itself, the power of the image comes from a small, insignificant dot barely discernable at the bottom right of the picture—our Earth.

Stunning almost beyond description, the power of this image comes from the way the sheer immensity of Saturn dwarfs the small,

Saturnian eclipse (Courtesy of NASA)

Detail of saturnian eclipse showing Earth (Courtesy of NASA)

relatively insignificant dot at the bottom right: our Earth. If one stops and imagines for a moment the feeling an astronaut might have in looking back from this perspective; the unimaginable beauty in stark contrast to the deep ache of loneliness of being 940 million miles from home.

This image brings to life for me the audacity and wonder of humanity's drive to explore, to always go farther. Like the ancient mariners who hopped on ships and sailed around the world with little expectation of making it back. Or other early wanderers who crossed the Bering Strait land bridge driven by a strong sense of "I bet there's something interesting over there!" Or those who invested the intelligence and resources to do something crazy, audacious and bold that required billions of dollars, such as sending a spacecraft on a seven-year journey to Saturn, a planet that Galileo first set eyes on in 1610. In the same spirit as these and other intrepid explorers from history, I believe we will carry on in a similar way in someday visiting Saturn. Once there, I have no doubt that we will gaze in awe at this magnificent celestial body and take a picture that says, "Wish you were here."

Cassini had initially been designed for a four-year mission. Nevertheless, year after year, it continued to transmit "I'm Not Dead Yet!"[32] and the mission kept being extended. Eventually though, its propellant gas tank was down to fumes from all the orbital maneuvers it had made over its 13 years at Saturn, and the decision was made to de-orbit the spacecraft. Without propellant, there was a remote chance of the spacecraft colliding with Enceledus, which could have contaminated a moon that might have life, with earthly DNA from a fingerprint missed during assembly.

Rendering of Cassini's Final Moments (courtesy of NASA)

---

[32] A tip of the hat to John, Graham, Terry, Terry, Eric, and Michael.

FROM THE GARAGE TO MARS

In September 2017, a command was sent for one last burn of the engines, slowing the spacecraft enough to cause it to fall into the Saturnian atmosphere. On September 15th, 2017, it entered the atmosphere, and a couple minutes later, it sent its final "beep." A last agonal breath before breaking into a thousand pieces and becoming a cometary streak across Saturn's sky. Our initials are still up there, reduced to plasma the atoms of our graffiti orbiting the planet, as it will be a million years from now.

At a space conference in the mid-90s, a bit before Cassini was launched, I had the chance to talk to an old friend, Max Benton, the CEO and founder of Able Engineering. Similar to our company in many ways but started ten years before Starsys, Max had formed Able to design and build large, deployable structures for spacecraft. These structures were a complex chaos of parts and pieces, and they were launched into space in a canister the size of a trash can. Once in orbit, the devices would open like origami and unfold into a structure the size of a football field. If you take a look at a picture of the International Space Station and its solar arrays, you will see sheets of Kapton held taut by long towers that look like they are made of Tinker Toys. That was Max's hardware.

Max was a hero to me. An individual with bucket-loads of experience in the space industry who was willing to take a call from me at any time and guide me through the many challenges of building a space company. His conversations with me were always filled with golden nuggets of great advice, and I had a file in my cabinet labeled "Max Wisdom" that was stuffed full of the notes I had taken during our calls.

At the conference, Max was clearly weary from the event and was wandering around the refreshments table, looking for anything to do but step back into the room full of hundreds of engineers listening to a speaker talking about the advantages of perfluorocarbon lubricants for space motors. We both now led successful companies, but in some ways, we were captains of ships passing in opposite directions. It was clear that he was ready to sell his company and retire, while I was full of piss and vinegar and was eager to shout "Bring it on!" embracing the next wild ride ahead.

During our conversation, I asked Max why he had done it. What was the payoff for investing his life in creating a space company? What was the reward that kept him pushing at the overwhelmingly difficult task of creating space stuff at the edge of what is possible and that absolutely had to work every time? He paused to reflect before answering in a simple, powerful way.

"That's easy," he said. "What makes it all worthwhile is that our stuff is enabling. The things we make enable humanity—changing discoveries that otherwise might not happen."

Max eventually did sell his company and retired a couple of years later, buying a house in Lake Tahoe and shifting his attention to teaching his grandchildren how to fish. Those last words from him turned out to be the final nugget of wisdom that he gave me, and his comment stuck with me because it went to the heart of why Starsys was so deeply rewarding for all of us.

Starsys did not build the cameras and telescopes that took the amazing pictures of Saturn. Nor did we build the spacecraft and rockets that brought Cassini to the planet. What we did was make the lens caps and latches and mechanisms that allowed it all to work, and if our stuff hadn't have done what NASA required of it, then the discoveries on Enceladus and pictures of Saturnian eclipses wouldn't have happened. Max was right.

The fuel behind all of the heartache and challenges and the pushing through what seemed impossible at the time was that we were enabling things that might not have worked without us. That is enough of a reason for me. Someday humans will go to Enceladus and we may find that life has evolved there as well. If so, I believe it will be stranger and more wonderful than we can possibly imagine, and those of us from Starsys will know that our stuff working flawlessly every time was our mark on that discovery.

* * * * *

A few years after we had put our mark on the Cassini hardware, we had a similar opportunity with an instrument going to Mars called MECA. My son Ryan was older then, and I had the chance to include him in the ceremony. One quiet evening after work, the two of us went to the clean room alone. We began by putting on dressing gowns and rubber gloves, and then with my hands around him, my body pressed to his in a hug from behind, I guided his nervous hands as he etched his signature on a piece of aluminum—the same piece of aluminum that landed on Mars four years later and that will be there centuries from now.

LUNCH WITH CARY

Revenue growth is entrepreneurial heroin.

ENTREPRENEURS AND THEIR companies are judged mainly by a single measure: revenues. More specifically, the company's annual sales volume[33] and how quickly this volume increases year after year. In short, achieve strong revenues that are growing robustly, and the value of the company is similarly strong.[34] What's more, be profitable while you grow and you'll achieve the valuation triple word score: strong revenues, growth, and profitability. Pull this off and the value of your company goes through the roof.

Those are heady times for an entrepreneur when revenues begin to ramp. When we acquired ATC in 2001, we had been struggling to maintain $6 million in annual revenues. It had looked then as if that was going to be the high-water mark with us settling into being a cozy 50-person company.

The addition of space motors to our quiver, though, changed everything. Once the industry realized we were able to deliver most of what moved on a spacecraft, they came to us with bigger and bigger challenges. Our track record of 100 percent success for our devices in space led to the conclusion in our customer's eyes that we could do no wrong.

---

[33] A company's yearly revenues (aka yearly sales) are the total value of what is sold during the year. The money the company makes every year is called simply "yearly earnings" (aka yearly profit). Another term used to measure a company's performance is EBITDA (Earnings Before Interest, Taxes, Depreciation and Amortization); pronounced "Ee-Bit-Dah." This is the raw earnings potential of a company, as the I, T, D and A can be tweaked and minimized by a smart accounting strategy.

[34] Interestingly, it is less the profitability (how much money you make) and more the revenues (how much did you sell) that determines a company's value. The premise is that if the revenues are strong then someone who purchases the company can always tweak company operations to make it more profitable. In the space business, the value of a company is typically 1 to 1.5 times its yearly revenue (this factor, called a revenue multiplier, is used as a rough benchmark of company value); but if this company is sustainably growing at 25%, the multiplier might be 2 or even 2.5. If it is a strategic acquisition it can put this ratio to 5 to 1 or higher.

FROM THE GARAGE TO MARS

Imagine yourself as I was: a CEO, founder, and entrepreneur owning 50 percent of a company with $6M in revenues, valued roughly at $6M. You are pleased you've created a good lifestyle company for yourself.[35] It's all goodness: a balanced life that works along with financial stability. You are the captain of a solid, thriving ship.

And then you bring in a new technology by acquisition and over the next two years, your revenues grow by 40 percent a year: $6M becomes $9M, which becomes $11M, and it looks to keep going. The revenue growth means that your company is now valued at two times revenues. Two years have passed, and your portion of the company is now worth $11M. You've earned $8M in two years by simply saying "yes" to the new business that was beating at your door. You are acknowledged by those around you for cracking the entrepreneurial code, and each year, you are given the opportunity to get even bigger. Growing another 25 percent next year means adding another $5M to the value of your portion of the company. Your response? *Bring it on!*

And your customer is complicit in the addiction. We revered our customer as all-knowing because, well—they were rocket scientists, after all. When NASA or Boeing or the European Space Agency says your company can do something, you believe them. Add to the mix the inherent narcissism and can-do optimism hard-wired into every entrepreneur, and the dialogue goes something like this:

> **NASA:** "We think your company is capable of building a docking system that couples two spacecraft and allows one to refuel the other."
>
> **Starsys:** "Are you sure?"
>
> **NASA:** "Yes."
>
> **Starsys:** "Well, all righty then!"

---

[35] The name for an entrepreneurial company created to provide a rewarding career and lucrative financial return to the founders/owners, in contrast to an entrepreneurial company created with the objective of providing a lucrative return to its investors.

LUNCH WITH CARY

We were on the heroin, and it became impossible for us to say "no" to new business. The past successes, coupled with the confidence our customers shared, had us living the dream—a *Mr. Toad's Wild Ride* of revenue growth with our heads out the window, wind blasting through our hair, whooping "Woooo Hoooo! Faster, Faster!"

We were spending time with astronauts, attending rocket launches, and helping design the stuff to go where no one had gone before. In only a couple years, we had gone from building latches and wax actuators to building refueling systems for spacecraft and deployable structures as long as a football field. Our revenues grew from $6M a year to $9M, to $11M, to $14M, and then to $18M.

We took on heroic programs, the ones with the toughest challenges that required our best talent to solve. And then we took on even more heroic programs. I was being asked to talk to space industry gurus, economic development leaders, and entrepreneur groups. I received many of the same questions: "How did you do it?" "How does it feel?" "What's next?" We became the go-to press opportunity for local media when something important happened in space.

As CEO, I was getting good at the humble brag. "It's really not me," I would say. "I'm fortunate to have great people around me." All the while my internal narcissist would be whispering in my ear, "Don't forget that you hired all of these great people."

As we brought one big program into the company after another, our technical directors suggested that we might not have enough resources to handle all the heroics at once. But the opportunities were always too good to pass up, and we were able to convince ourselves that we would figure it out as we always had. We had a single dab of peanut butter to cover a piece of toast that was getting bigger and bigger, and we told ourselves that all we needed to do was spread it a bit thinner.

FROM THE GARAGE TO MARS

The signs that the wheels were starting to come off were subtle at first. Here and there, different programs began reporting problems; while they were solvable, they seemed to be popping up more frequently. Each of these problems gnawed at our profitability, but there were always the winners to balance out the losers, as we had many programs under way under their budget at any one time. The trend of the losers beginning to outweigh the winners every month was easy to ignore when caught up in the excitement of the next big program that had just been won.

* * * * *

On a Honeycrisp-cool October day in 2004—the same year that we would reach $18M in sales—Cary Ludkte called and asked me to lunch. Cary was head of the commercial spacecraft division of Ball Aerospace, the 1,000-plus person aerospace company in Boulder that had been a key customer for us during our formation. He was keenly aware of our wild growth, as we were building important hardware for his division. I expected he might be wondering how we were handling our success.

The signs that our space company was getting in over its head had started to manifest themselves in our internal financial reviews, creating a dull thrumming of subconscious dread. This first started showing up in our Strat Team, our leadership strategy team that met once a month to look at each program in detail. Made up of the heads of Engineering, Operations, HR, Marketing, Quality, and myself, it was our version of Mission Control, where we looked at the health of each program each month. The meeting discussions had been getting longer as we worried how we would right the upside-down programs. Usually, these conversations were focused on only one or two programs, but now they were stretching to a handful of discussions. The situation was beginning to concern us all at a subconscious level, but each month, we figured that next month's numbers would improve as we kicked the can down the road.

I swept any of this anxiety under the rug as we sat down for lunch. I was excited that Cary had something to share that warranted lunch at Le Chanticleer.

I had earlier in the year floated the idea that Cary consider leaving his role at Ball to take a leadership position at Starsys. For certain types of executives, the siren call of a start-up resonates with their desire and ability to create something of great significance. For this reason alone, Starsys seemed like an especially attractive opportunity for Cary.

After drinks were served and our order taken, he pulled a single sheet of paper from a folder, set it on the table, took a brief final look, turned it 180 degrees, and slid it across to me. The ceremony of this was not lost on me. He had put much thought into what was on the page, and he was treating it with care.

"I've been thinking. I would like to join you in leading the Starsys team," Cary said. "I talked it over with my wife and we're both on board. I think the combination of your vision and my experience and connections could take Starsys to another level altogether. Here are the terms that could make that work."

The document was simple. A few bullets explaining his salary and benefit requirements, as well as a bonus in the form of company stock if he succeeded as he expected. He also included a couple of paragraphs on what he proposed his responsibilities would be that would play to his strengths.

I went silent as I took a couple of minutes to read the document. The first thing that hit me was that what he was asking for was more than reasonable—it was a bargain. If I had spent $200K on a nationwide search for executive talent, I would likely not have found someone as qualified for the role as Cary, and, if I did, I would be paying a salary twice what he was proposing. I was also struck by the sincerity of his desire to be a part of and to contribute to Starsys—and by his belief in himself that he would succeed greatly and be rewarded as a result.

The second thing that happened as I read his proposed responsibilities was a barely conscious whisper of, "But what would I do?"

Much of what he was proposing—including leading the development of customer relationships and overseeing the operations of the company—were things that I did. It wasn't that I was worried he wouldn't succeed in these things, as he was a great talent I expected would succeed at whatever he pointed that talent at.

Instead, I had another concern. I was worried that he might be better than me. Entrepreneurs often talk about the success of their companies as being paramount, but at the same time, they harbor an attachment to that success as being a direct result of their leadership. What if a different leader would be better for the company? Or what if a co-leader eliminated the majority of responsibilities of the entrepreneur? "But what would I do?" was the self-talk that repeatedly bubbled under the surface as I read the proposal, and it influenced my response.

"Cary, this is great, but I'm not sure what you're suggesting is what we need right now," I said, sliding the paper back. "It's a bit redundant and overlaps with some of the things that I'm handling. I'm looking for a partner who is more focused on strictly operating the business." This was true, as I had been looking for someone to complement my role, but not overlap with it. However, what was also true at that moment was the emergence of a strange feeling that I could sense but not articulate—that somehow at that moment something important had happened—with both Cary's offer and my decision to pass on it.

Cary had gone out on a limb with his proposal. In one sense, it was crazy for him to be considering leaving a leadership role in a major aerospace company for a start-up that might not ever succeed

LUNCH WITH CARY

and a CEO role that might never materialize. He paused for a couple beats, then:

"No problem," he said. "I wasn't sure the time was right but wanted to give it a shot." He took the sheet back and tucked it into his folder, as if the offer were being officially withdrawn. There was no "I'll leave this with you and mull it over."

We poked at the idea from a couple of angles during the rest of our time together. However, we decided in the end that it was best to hold off on any further discussion and to revisit the idea if the sun, moon, and stars were to align in the future. In coming to this conclusion, though, I had an increasing sense of unease, realizing that there was possibly a great window of opportunity here—and that I was closing it for good.

I have a near-perfect, flash-bulb memory of pushing through the door of Le Chanticleer into the afternoon after our lunch, cumulus clouds bunching in the blue sky above, offering a hint of the afternoon thunderstorms that were a couple hours away. The close of the door behind me underscored the feeling of dread as I began to realize what had just happened.

In that moment, I recalled that when Cary had passed over the proposal and as I first started reading, I had felt the flush of goose flesh suggesting another Nudge: a serendipitous, improbable opportunity showing up out of the blue.

As I walked toward my car that October afternoon, I understood that I had said "no" to the Nudge, my first recollection of ever having done so. The Nudge-whisper of, "But just maybe?" was overruled by the voice saying, "But what would I do?"

I swallowed the feeling that I had just said "no thanks" to the one individual who could navigate us through the storm I felt was coming, hopped in my car, and with a wave, drove back to the company.

FROM THE GARAGE TO MARS

Tibbitts Tip
Founder's Syndrome

More than 80 percent of technical entrepreneurs are poorly equipped to grow a company beyond a certain size. Not from lack of experience, but because the wiring that creates people who can create something from nothing, is very different than that needed to scale what has been created. For some founders, the ceiling is 10 employees, for others, 100. Rarely is this thousands (Steve Jobs and Bill Gates being notable exceptions).

My personal ceiling (in retrospect) was in the 75- to 100-employee range. This is an extraordinarily difficult thing for the Founder/CEO to understand and recognize, as by definition a founder is someone who can do it all. Several things conspire to make this so:

- A powerful attachment to the title of CEO; the belief that if you aren't the CEO, you are no longer a rock star.

- A belief that as CEO, you have to do it all. That's what a CEO does.

- The belief that the CEO has a certain way of doing things that is the reason the company is successful, that will be lost if someone else has the wheel.

- Ignorance of the magic that happens when you do find the partner who loves the stuff you hate and is a wizard at what you struggle with. Usually, these are operational responsibilities. A COO is often a powerful partner to a visionary CEO. Sometimes it is needing a business development rock star, although that is often a later need, as sales are often a strong suit of the founder.

    Jokingly, this is referred to in the investment community as the Founder needing "adult supervision." This actually is said with high respect. It's a way of saying "For God sakes, you've done the impossible. You've created a company from scratch. Now find the leadership partner you can trust who has successfully taken companies to the next level, so you can relax and continue to be visionary."

- The importance of this cannot be overstated. It is the elephant in the room that prevents a company from going big. And this needs to be done proactively, when you start to feel a wobble in the car, not after a wheel falls off.

- And...as Yoda would say, "Choose carefully your partner you must."

    (See the next chapter for what happens when you choose that partner from desperation.)

    If I had said "yes" to Cary, Starsys would have had a very different future.

    An excellent treatment of founder's syndrome and a cure can be found in the book *Traction* by Gino Wickman.

THROWING HATS OVER FENCES

THERE IS A STORY TOLD OF a man walking along a country lane in Ireland. The lane was not one for getting from here to there quickly; rather, it curved its way with lazy abandon through a landscape of undulating, emerald-green fields with an occasional stone fence demarking whose land belonged to whom. He was headed to a small inn that he knew of further up the lane and as he came around the corner, he could see it in the distance. His pace quickened toward the warm food and cold pint that he was looking forward to enjoying.

Earlier on the journey, as the road had meandered in a direction away from the inn, he had considered taking a more direct route. Although he knew which way the crow would fly, he thought better, knowing that if he came to a stream or fence, he would have to back-track, which would double the time to get to the inn. He was too hungry to take the chance.

As he rounded the next bend, he could see the inn clearly about a half-mile ahead. He had been right to choose the road. There was a tall stone fence he could now see that stretched for at least a mile across the fields, which would have spoiled the shortcut.

Looking across the fields, he was surprised to see a well-dressed gentleman in a suit and bowler hat taking the same short cut that he had considered. The gentleman had not yet come over a final rise and was unaware of the fence ahead. Despite his hunger, the man paused to see what would unfold as the gentleman came over the rise and realized he had a fence ahead of him that he would be unable to cross.

FROM THE GARAGE TO MARS

As he came up over the rise, the gentleman paused when he saw the fence. After a moment of study, he approached it to find a way over. He walked back and forth along the fence for several minutes, trying at various locations to climb up and over the stones to the other side. With each try, a foot or hand would slip, and he would be forced to look for another route. After several minutes of attempts, it was clear he was not having any luck.

Then, as the man watched, the gentlemen took off his hat, looked at it for a couple of seconds, and threw it over the fence. The gentleman then continued his hunt for a way over. After a couple of additional failed attempts, he found a spot where he was able to put his feet and hands just so and pull himself up and over. He succeeded in landing on the other side, picked up his hat, dusted it off, and carried on.

The man was fascinated and puzzled by what he had just seen. With his curiosity piqued, he picked up his pace so that he would rendezvous with the gentleman in just a short distance.

After meeting a few minutes later and exchanging introductions, the gentleman shared that he was on his way to a reception. Then the man asked his question.

"Please excuse my curiosity. As you crossed the field and came to the stone fence, I couldn't help but notice that you threw your hat over the fence and then found your way over."

A broad smile creased the gentleman's face. "Ahhh, you saw that, did you? For the life of me, I wasn't finding a way over that cursed fence, and I certainly wasn't of the mind to turn back to the lane and be late for the reception."

"This bowler is irreplaceable," the gentleman said, continuing his explanation. "It was given to me by my father, who it was given to by his father. I threw the hat over the fence because there was no way I would lose something that precious. When I threw it, I had no choice but to find my way over!"

THROWING HATS OVER FENCES

If there was a parable that Starsys lived and breathed, that was it. We had survived birth and flourished by throwing hat after hat over fence after fence. It is what we did, what we were known for. We had always delivered, no matter what the challenge. We made bold commitments to our customers that we would find a way and always had.

We had by now deeply ingrained "it had to work in space no matter what" into our corporate DNA. Our confidence in our ability to do the near-impossible had created a "Bring it on!" bravado that both our customers and our company thrived on.

The more we delivered the near-impossible, the more opportunities came our way. The more heroic the programs we delivered, the more we convinced ourselves we could do anything.

Our success created a formula that was unsustainable and at some point, something would have to give. In retrospect, the path we were headed down required the moderating influence of seasoned aerospace leadership to rein in our appetite for challenge to a manageable level. Exactly the skill set that Cary Ludtke would have brought to our company, and whom I had said "No" to at the Le Chanticleer.

In most companies that grow too quickly, the cracks in the dam are compromised quality. Too much to do leads to products that don't work as promised. As word gets around, growth stagnates giving the company time to sort through its problems if it can. It's a corporate Peter Principle[36] that ensured we would eventually get in over our heads. However, we had a different, more insidious problem: We had so deeply ingrained quality into our culture that

---

[36] A popular theory of organizational leadership published in 1969 by Laurence Peter and Raymond Hull states that competent individuals continue to be promoted within an organization until they are finally promoted into a role that they are incompetent at, whereupon they become "parked." Taken to it its limit, it ensures that any organization will be managed by individuals who are incompetent in their leadership roles. A generalized rule of the Peter Principle, applicable to Starsys, is that a system that works well will be used in progressively more difficult situations until it fails.

delivering substandard product was not an option—the only degree of freedom left was the money we made.

With a commitment to never deliver anything less than extraordinary products, profitability became the casualty. At first, there were just a few leaks in the ship. Profit numbers would be lower than expected on a particular program but could be made up by heroics on other programs. We told ourselves that these instances were only temporary downturns, a coincidence of simultaneous bad luck on certain programs. We were able to convince ourselves that we were simply taking a temporary hit to profitability; it was nothing we couldn't recover from. After all, we had been profitable every year of our existence.

We had galvanized the company around the collective mindset of *"it absolutely has to work in space, no matter what it takes."* With that brand ethos seared into our DNA, we eventually threw one hat too many over the fence, and later too many all at once. In short, what gave was our ability to make a profit. The programs began to lose money as we began to make mistakes, and the problems were compounded by our efforts to correct these mistakes to make sure that nothing we delivered would ever fail in space.

What was brewing here was a perfect storm of a magnitude that we could not appreciate—with the final factor our obtaining of bank financing. To best understand this, allow me to offer a quick tutorial on growth capital.

A company that is growing rapidly usually needs cash beyond the money it is earning. In our case, we did not get paid for the work we did for about 90 days after we completed a given project. We would typically build a piece of spacecraft hardware, pay our employees for the effort, and then three months later, receive payment from our customer. When you are in fast-growth mode, your need for cash outstrips the money that you are able to retain from profits. Our monthly revenue rate was now $1.5M per month, which meant that we needed $4.5M in extra cash to pay for our employees work for the three months prior to that work being paid for by our customers.

There are three paths to growth capital. If you are growing moderately, you can provide that cash through profits that you have made in the past. For instance, if you've accumulated $2M in profit through the past years and need only $1.5M in additional operating cash, you can self-finance the growth with the profits. You are beholden to no one for the growth capital.

But when growing quickly, you need more money. Sometimes lots more. The good news is if you are growing quickly, there are two groups willing to provide those monies to you:

1) Banks, notoriously conservative, nevertheless are willing to loan money to you at a reasonable rate of maybe 7% a year when they see a strong track record.
2) Investors, who are willing to give you money for stock in your company despite the risks.

The decision which to take often hinges on the cost of the money. If your company is growing at 40%, as ours was, and you sell stock in your company, the next year that stock is worth 40% more and the following year, another 40%. Essentially, you are paying 40 percent interest on investor monies.

For example, let's say we need $1M in growth capital. If I get that money from a bank that is charging 7% interest, that $1M costs me $70,000 for the year. But if I instead I sell stock for $1M to get that money, the next year that stock is worth $1.4M because the company grew 40%. I will have to pay that shareholder $1.4M when I sell the company, essentially paying $400,000 for that same $1M. When looked at through this lens only, bank financing is a bargain for growth capital.

FROM THE GARAGE TO MARS

Our success had triggered a desire of one of our Japanese partners to invest in our company. They were asking to purchase $3M in stock in our company as a strategic investment at a stock price that was more than fair. But we could not get past the calculus that we would be paying an additional 40% to those investors for every year we grew 40%, which raised a critical question: Wouldn't it be better to get the money from a bank that would charge us only 7% interest?

And the banks were more than willing. Because of our track record of 40 straight quarters of profitability, it was easy for the banks to convince themselves that Starsys was a sure bet.[37] Our successes had been so convincing that we were being fought over by two banks wanting to finance our growth. The cheap money was too good to pass up, and so we said "no" to our Japanese partner's investment and doubled down with the bank, eventually closing a $6M loan at a low interest rate with the ostensible guarantee that we would have all the cash we would need to continue our growth in the coming years.

What we were not including in our calculus, however, was what these partnerships morph into if things start to go south. When your money is coming from someone who has become a partner in your company, a shareholder who is now fully on board wanting it to succeed greatly, you have an ally, often with deep pockets, willing to help bail in heavy seas.

Not so with a bank. The conservatism of a bank and its low lending rate frequently leads to "hooks" in the loan that ensure that the bank will get its money back—no matter what. These hooks appear in the form of personal guarantees from the CEO[38], as well

---

[37] Things have changed. That was 2005—before the sub-prime loan debacle created the oxymoron "easy bank loans."

[38] When Starsys was started and we received a first modest $50K loan from a bank, it required a personal guarantee from the CEO (myself) that stated that if the company went bankrupt, I personally would guarantee the loan was paid back. An easy give when my only significant asset was a 1980 Ford Fiesta with 110,000 miles. But personal guarantees are like a bad tattoo impossible to get rid of once in place. A personal guarantee ensures that if the company sinks, the captain goes down with the ship, exactly the motivation the bank needs to commit their dollars. Every entrepreneur I know has made one personal guarantee on a loan but never a second; the first experience creates a mantra for the rest of their lives of "I will never again personally guarantee a loan."

as loan covenants, which basically state in effect that "we are pleased to loan you our money as long as your profitability continues. However, should this change and your financial status change, we would like our money back immediately, thank you." This is the corporate equivalent of the right to foreclose on your home mortgage simply because it looks like you might not be able to make your payment.

When you are borrowing $6M, you agree to covenants, which means that all you need is a hiccup in your financials and the bank can call the loan. It is like having a home loan and the bank having the ability to foreclose if your boss gives you a pay cut. In the case with our loan, the covenants were tied to our profitability. If we had three months of significant losses, we would technically be in violation of our loan covenants, and the bank could call our loan. In general, "call the loan" means that the bank can demand full repayment of your loan and they can assume ownership of your company if you are unable to pay the money back. In our situation, this had been a sure bet for the bank as our loan was $6M and our company was valued at $20M, which was the reason for the bank's comfort in loaning us the money at 7%. We went into the loan with our eyes wide open, both the bank and Starsys confident that a streak of 40 previous quarters of profitability ensured smooth sailing.

The last piece of the perfect storm was in place. It was now just a matter of time.

ENTREPRENEURIAL HELL

IT WAS 2005 AND I WAS taking a respite from the challenges of the company, enjoying myself in a video studio in Denver. We had received a $100,000 grant from the State of Colorado to create a training video for space technicians, the state recognizing that we were creating high-paying tech jobs by hiring motivated high school graduates and helping them to become spacecraft makers. The set had a green screen, multiple cameras, a teleprompter, and an actress we had hired who was helping me explain the "10 Rules of Making Spacecraft Hardware" to the camera. Although in a sense it was an odd time—given the company's challenges—to be creating a video feature on the ins and outs of making our hardware, it was nevertheless a wonderful break from the worry of multiple programs that were struggling.

During a break, my cell phone rang. I stepped into a darkened soundproof room next to the brightly lit set, watching through the window as the production team repositioned the cameras for the next scene. Although this room was completely quiet, I was surprised by the strange sound that emanated from its silence, as if cotton had been stuffed in my ears. I took the call. It was like an oncologist calling and leading off with "I'm so sorry..."

"Scotty, bad news. We were ready to ship the electronics module for the Deep Impact spacecraft and we were going through testing that included a final high temperature test of the hardware in an oven last night. However, the technician didn't monitor the temperature properly. The oven was a hundred degrees too hot, and the electronics melted."

You're kidding me, right?"

"No."

"Do we have time to remake the electronics to support the schedule?"

"Yeah, I talked to the customer. They can work with us on it, but that's not the worst of it. It will cost us half a million dollars to remake the electronics, and that will have to come out of our pocket."

"Any way to get the customer to pay for it? Any insurance?"

"Nope. It's our fault. We have to cover it."

I paused and let the news sink in before responding.

"OK. We'll figure it out. Go ahead and get the parts on order. Let the customer know we'll make good. How's the tech doing?"

"Not well. He's besides himself for making the mistake. He's sure he's going to be fired. He feels terrible about doing this to the company."

"I'll talk with him tomorrow and let him know I did the same thing 15 years ago. This stuff happens when we're going fast. Let him know."

"Will do ... hey, Scotty, I'm sorry. This one is going to be tough."

"Yeah, I know, but we'll figure it out. We always do. See you tomorrow."

I walked out of the soundproof room and back onto the set. The transition from dark to the bright lights of the set was a bit of welcome relief as I found myself putting my sense of dread aside to finish the shoot. Later, driving home to Boulder, I faced the reality that we had likely reached the tipping point and were headed into very rough waters. It now seemed inevitable that we were going to crash. That we had thrown one hat too many over the fence. The only question now was how bad would it be and how would we get through it? I truly had no idea what we would do other than to wait as it played out and react as well as we could.

There was no longer a way to look at the financial performance of the company and wiggle our way through to showing profitability. When you recognize that program losses are real and not going to be recovered, you must present the losses in the monthly financial report and send it to the bank for review as required by the loan covenants. In reviewing our program losses, we were now showing a $1M loss for the company quarter.

In our case, the response was immediate. The bank called a meeting to go through the numbers and confirm the improbable: After 50 consecutive quarters of profitability, we were now losing money quickly.

I sat at the head of the table in our conference room with a handful of top execs from the bank as director of programs and chief financial officer explained that we were now in a perfect storm of troubled programs. I looked around the room and saw that the bank representatives who had underwritten the loan were as nervous as we were. They were the ones who had assured their leadership that we were a solid bet for a multimillion-dollar loan.

We went through each troubled program, explaining the challenges, all under the umbrella of "spacecraft hardware building is a hard business." This they understood. What did not make sense was why multiple programs were struggling simultaneously. It wasn't that any one program by itself was overrunning its budget. This had happened in the past when one struggling program's losses had been more than made up for by the dozens of programs that were on track. The excruciating problem now was that multiple programs were simultaneously bleeding out. The balance of programs was still profitable, some significantly so, but in a business of thin margins, the several programs that were losing money were like an anchor attached to a swimmer treading water. The weight of the losses were pulling us under. [39]

Loan documents are the poster child for "read the small print." They include dozens of covenants that explain the relationship between the bank and the company in covering what the rules of the loan are. These are so boiler-plated that they really aren't documents that are negotiated; they are presented by the bank with a context of "if you want our loan, this is what you are signing up for." Threaded throughout the document are the requirements the company must meet to be in good standing with the bank. If the company's financial performance falls outside of the covenants, the bank has the option to increase the interest rate while simultaneously calling the loan, basically saying, "If your financials get flaky, we will demand our money back, or we will take the collateral." In this case, the collateral was our company—every asset, every patent, every desk, every vacuum chamber, and every clean bench. All of our current contracts and all of our future business. All would be transferred to the bank if it called the loan, and we were unable to pay.

---

[39] Most spacecraft programs are run as "cost plus" programs, meaning that the customer, for instance NASA, pays for the costs of the program, as well as a small profit (the "plus") on top of the contract cost. The profit for these types of programs is typically 6% to 8% ... the government figuring that you should make only a small profit if they are going to guarantee you don't lose money by paying for your overruns. If you spend significantly more than you originally budgeted, the customer will pay you for those extra costs, but it may reduce the profit you make from 8% to 5% as a penalty ...but ... you always get paid for the work you do.

Another factor in Starsys' success had been that we did the unthinkable: We promised to do a spacecraft program for a Firm Fixed Price or FFP. This was a godsend for our customers, a group constantly hammered by Congress for their constant overruns, as it meant that price for spacecraft (or at least the Starsys part of it) could be capped. It was a ballsy way to do a contract, but if you had the courage and competency, you could make significantly more than 8% by doing it well. Some of our more successful programs made as much as 50% profit, but it took superhuman program managers who could maintain a budget even when things were going pear-shaped on the program. Few FFP programs made these large profits, and they were capable of huge losses. If your hardware fails during testing, (because something wasn't designed just right ...which is what happens when you have too many programs and not enough senior engineers to oversee them all), you might have to spend twice as much as the contract, as you have to make the hardware twice and take twice as long to be paid. This is the situation we found ourselves in at this point, as we were carrying 50 programs, 40 of which were successful with small profits and 5 to 10 challenged with significant losses, all from small errors such as the oven over-temperature. It was like a ship with too much weight on the top and not enough ballast on the bottom. A wave hits it, and it flips.

ENTREPRENEURIAL HELL

And buried within the document was the personal guarantee of the CEO, me. Every asset that Jackie and I owned—our house, savings, furniture, cars—everything was included in the collateral. If the bank called the loan and the company was unable to pay, all of that would be theirs to sell to recover the $6M that was owed.

After hearing the news and the projections for the next months, the bank execs left, seemingly conveying at a surface level the sentiment, "These things happen, but we'll work together to figure it out." But their look as they left and shook our hands reflected the truth. Their faces evincing sadness and pity rather than confidence. They knew what was coming. We could only imagine.

During the following week, we received a few calls from the bank asking for more clarification, and then another meeting was set up to discuss the path forward.

At the next meeting, the bank said it had decided not to take ownership of the company immediately. Rather, it explained we would be given the opportunity to sign a forbearance agreement. The forbearance agreement stated that in exchange for our company agreeing to not declare bankruptcy, the bank would provide a certain amount of time for us to pay back the loan.

During this period, the interest rate on the loan would be doubled. We would be given five months to raise the $6M required to pay back the loan and then we would be back in satisfactory standing. As an act of good faith, though, we would have to pay $1M in 60 days. If we failed to make the initial payment, or after that, the final payment, the bank would chain the doors of the company and take ownership. In addition, it would simultaneously take ownership of my personal assets. All would be sold to pay off the loan.

We have all seen movies in which people somehow get sideways with a debt to a loan shark and must pay back an impossibly large sum in a ridiculously short amount of time or else lose their knees, a limb, or their lives. There is a reason this trope is used so often, as it elicits a powerful, sympathetic angst for someone given a Herculean task as a result of having done wrong and who will be punished and suffer great pain if the task is not accomplished. I've always viewed these situations with self-righteous hubris, believing that I would never get myself into this kind of a predicament and thanking God that it could never happen to me.

Yet that was the situation I now found myself living. It was mind-numbingly frightening. I remember sitting at my son's Little League game shortly after the meeting, not able to think of anything but what was ahead. While my wife would comment on the game and our friends would be catching up with each other, my mind kept repeating, over and over, with no resolution whatsoever, "*What am I going to do? What's going to happen?*" I could do nothing but picture losing the company, my home, my family—the family I loved and had committed to providing for—wondering if they would stick with me after the harm that it appeared I was about to cause them.

The summer of 2005 became hell, as I was working to navigate impossibly rough waters while terrified of what was likely to happen if this effort failed. I began the process of trying to find investors interested in our company, wondering if it would be possible to sell off just a portion of the company to raise the money that was required.

Prior to that year, it would not have been difficult to value the company at $15M or more; the $6M requiring us to give up a little more than a third of the company for the resources to pay back the loan. But we were now what banks called a "distressed company." With Starsys hemorrhaging money, it was now worth a fraction of what it had been a year ago. It was unlikely we would be able to raise the $6M without selling the company outright.

ENTREPRENEURIAL HELL

It was at this time that the bank introduced us to a talented investment banker who was head of an investment banking firm in Denver our bank felt might be able to help with the process.[40] What I did not realize that our meeting was also a test. If he walked away unsure we would be able to sell our company, the bank would take an even more aggressive stance with regard to getting their money back.

Our first meeting started poorly. The investment banker clearly conveyed the impression that he was doing a favor for the bank by seeing us, and not believing we would be a solid enough proposition to craft a sale for. As we told our story, I remember the moment the lights went on for the banker. He leaned in suddenly, realizing that although there were significant cracks in the ship, overall it was sound and capable of great things—if only the cracks could be patched in time. By the end of the meeting, he was willing to give it a go and we signed a contract that paid him a portion of the sale if we were able to pull it off.

If we could get $6M for the company, I would get out alive, keep the house and the family, but would lose everything that had been invested in the company over the past 20 years. The portion of the company I owned, that had been worth millions a year before, would be worthless. I either needed to find someone willing to invest $6M for more than half of the company, or else sell the company outright. The shareholders of the company would receive anything left over after the loan was paid back.

Looking for investors or buyers of the company was like turning over an hourglass with five months of sand in it, knowing that we had to find the investors before the sand ran out. Selling a company is usually a multi-year process. We only had months.

---

[40] An investment banker packages businesses for sale, looks for buyers, and shares in the proceeds. Essentially, an investment banker is to selling your company what a realtor is to selling your house.

FROM THE GARAGE TO MARS

Although I was getting advice from all corners, the fear, mind-chatter, and second guessing were so loud that I was unable to sort the truth from the fiction. Some said that I was on the brink of losing my company. Others said that the bank would never chain the doors; that they would threaten but would continue to work with us even if we didn't make the payments. The second story provided comfort, and so in my state of survival, I adopted that as my truth. "Maybe I really do have more time," I remember thinking.

While I was working through a list of potential investors who had been pounding on our doors the previous year but who were now treating us as damaged goods, what I did not know was that my chief operating officer, Travis Phillips,[41] had gone to the bank during this same time without my knowledge and suggested another plan. When I found out about this, my mind was quick to imagine how Travis might have played this: "You know entrepreneurs ... Scotty's heart is in the right place and he's a good man, but he just doesn't have the experience to make it through this. Fortunately, I've been in this business for decades and can see a clear path forward ..."

Travis was a highly experienced space industry executive with an impressive resume. I later learned that he told the bank that he had seen the problems coming but was unable to convince me to make the necessary changes, a clear tell that I was not qualified to lead the company through the storm. He conveyed that if I were relieved as CEO, he was fully capable of turning the company around in six months given the chance. He had presented his plan for recovery and explained that after fixing the problems, the

---

[41] Name has been changed. Travis had joined the company several months earlier, when the wheels had started to come off the cart. I had realized my mistake of not hiring Cary. Cary was no longer available, and I was looking for similarly talented leadership help. Travis had heard of the need, chased me down, presented as a similarly talented leader, and I eagerly hired him, vowing to not make the mistake I had prior with Cary.

company could then be sold at a premium, recovering the bank's money and more. He would be willing to take on the CEO role if given a significant equity stake in the company.

In essence, Travis was presenting a path forward that would get the bank out from under the loan, and it would also allow the bank to carry out a smooth transition after initially chaining the doors. Once the bank pulled the trigger on this plan and took possession of Starsys, the mechanics were such that the employees would arrive at the company the following day and would not be let in the door.

Travis would then reorganize the company according to a plan he would have laid out prior to the repossession. I would be fired, and the bank would open the doors again for business a couple of days later. In this new organization, the bank would own the majority of the company and Travis would be the new CEO with a significant equity stake.

As it turned out, I did not learn about these discussions until long after our coming investment deal had closed.

The advice I got that the bank would not chain the doors did not take into account that I had someone on my executive staff who had promised to helm the ship through the rough waters if the captain was relieved of duty.[42] My understanding is that the bank carefully considered this offer, as it gave it the best of many worlds including a path to untangle all with someone at the helm who was "their guy."

---

[42] Shortly after I hired Travis, I phoned a senior aerospace executive who had worked with Travis in the past to give him the great news that I had hired Travis to help us through our growth challenges. The executive had known Travis well from previous work in the aerospace business. There was a long pause, followed by a measured, "Well, if you hook a cart to that horse, it will move. Good luck." Four years later, the executive shared with me that after he hung up the phone, he had put his head in his hands and said to himself, "Crap, Scotty just lost his company."

FROM THE GARAGE TO MARS

In the end, though, the bank said, "No, let's give Scott a chance." I found out later that the tipping point in my direction came from a decision based on one of its core values: integrity. For the bank, it wouldn't be right to not give the CEO a chance to fix the problem he had created. At the same time, the bank made it clear to Travis to stand by. If Scotty didn't meet the requirements of the forbearance agreement, it would quickly pull the trigger on Plan B.

Inputs and advice came from all directions, making it nearly impossible to sort out what to and what not to do. One individual suggested that if the house was in Jackie's name rather than mine, it would be protected from the personal guarantee. As I prepared the paperwork for the transfer, another attorney informed me that in our situation, the transfer would be considered fraudulent and therefore annulled. I threw out the paperwork, realizing the house could not be taken off the table.

I received reports that an "unnamed entity" was looking into our assets to confirm their value. Chillingly, in the background, the bank was doing its diligence, calculating how much it would be able to recover if it repossessed our assets.

I was making calls to the companies that had previously expressed interest in investing in or purchasing Starsys. We did our best to present the company as sound but simply going through rough waters. We had interest at first. Every meeting started with the hope that "this is the one that will see the value despite the losses." However, as each prospect visited and looked closely at our company, each one invariably sensed the reality of our situation, and it either suggested a valuation that could not work for us or passed on the opportunity.

During these meetings, we found it nearly impossible to keep the panic of the situation from leaking into the room. In the middle of one meeting with a promising investor, our CFO cracked,

ENTREPRENEURIAL HELL

emotionally stating through tears, "You've got to invest in us. We don't want to lay off all these people!" These are the kind of statements that turned off interest like a light switch. Imagine being out for dinner and over dessert, your date blurts out, "You don't have a problem with chronic herpes, do you?" The investor politely but immediately passed.

By June of 2005, four or five companies had taken a look at Starsys without any significant interest. It was becoming a pattern: initial interest and surprise that our company was looking for a buyer or investor. Next, there would be an enthusiastic visit, during which the level of our distress would become apparent. This would result in a quick exit, and a follow-up "thank you for the time, but we are not interested" call.

I was not sleeping. Going to sleep was next to impossible with all the brain chatter. I would wake at 2 or 3 a.m. with a shot of adrenaline as some random dream hinted at losing it all. My thoughts would then go into overdrive trying to solve an unsolvable problem. I would look at the clock ... "OK, it's 3 a.m. ... if I can get to sleep now, I can get three more hours and be OK ... OK, it's 4 a.m. ... if I can get to sleep now, I can get two more hours and that will suffice ... OK, it's 5 a.m. ..." and on it went. I was not able to function properly on the three to four hours of sleep I was getting.

Jackie did everything she could to help. She gave me cards that reminded me of the good husband and father I was. She put photos of the family into frames and handed them to me in gift bags. I surrounded myself with these on my desk—a cocoon of pictures trying to protect me from the hurts of each day. I thought it was a subtle thing that others wouldn't notice, but those who walked past my office couldn't help but see what was akin to me trying to wrap myself in a comfort blanket.

I lost my appetite and wasn't eating. My weight dropped 15 pounds and my clothes hung loosely. There were a few advantages.

FROM THE GARAGE TO MARS

I remember saying with a nervous chuckle that I could go to Red Lobster and eat as many of the cheese biscuits as I wanted without remorse.

I found out I was not suicidal. That that I was not an alcoholic. I had all the motivations for both. I would come home and curl up on the couch and cry, working to get it out of my system so that I could go in and put on a game face the next day. The proud, confident leader whom my family had always known, the rock star entrepreneur, was gone.

I discovered that I was not alone. Stories started to surface of other entrepreneurs who were on the path to losing their companies, and their loss of ability to lead. That provided some comfort.

My daughter, at the time seven years old, sheepishly walked up to me one day.

"Dad ... Mom said your company needs money, and you're sad because you don't have the money you need. I've been saving mine up from birthday gifts and have $55 dollars. If I gave you half of it, would that help?"

Tears welled. "No, honey ... thank you so much, but I need a bit more than that."

I remember being on a rooftop deck at a restaurant with 20 good friends, having drinks, laughing, telling stories. As I sat at the corner of the table watching with the sun warming us and the mountains behind Boulder as a backdrop, I was completely and totally incapable of participating in the fun. Looking around the table, I was envious of all of these people who were able to leave thoughts of work behind, who could laugh and enjoy life—a stark contrast to my experience more than 15 years earlier of walking into a health club believing there was not a soul in that building enjoying life more than I was.

For years, my wife had been gently asking me to give my life to Christ. I had been raised Presbyterian, and considered myself a

Christian, but she had been asking me to surrender myself at a deeper level. One afternoon in early May, we were leaning against a chain-link fence, watching baseball, and she spoke:

"I know you don't want to surrender this problem to the Lord. That's your choice and I respect that. But if you don't have a deep level of faith to support you, I believe that you will likely not recover from this, and you will personally crash. That's your choice, but realize that if you do, you take all of us with you."

After all the years of her asking me to have a deeper relationship with God, this was the one that got through. This shook me enough to reach straight to my heart and break it. That night, we invited our pastor over. I took communion and two weeks later, I was baptized a second time, immersing myself in the water, like a child floating in the warm water of a womb, nostrils pinched by fingers, a prayer on my closed lips: *"Lord ... you've got to get us through this."*

Even if not always apparent, I had begun to realize that I was not in this alone, as there was now high cover from my faith, which helped me sleep and helped guide decisions. Throughout the coming months, faith played an important part in working our way through what lay ahead, providing some measure of solace and peace as things got worse before they got better.

The first deadline was coming up: June 15th. I was to deliver a $1M check to the bank, and I had no way to do so. There were some possibilities that were beginning to emerge that looked positive but were going to take time to develop. One deal that was starting to come together involved a group of investors from the space industry; friends and colleagues who would each invest a million dollars or so. In reality, though, these deals were far from a lock and were simply not going to close anywhere near the June time frame.

FROM THE GARAGE TO MARS

The deadline for the first million dollars was looming, and our family needed a break from the crazy-making stress, so we headed to southwest Colorado for a week at the Wildnernest Trails guest ranch. We were far from Boulder, and every day, we would take horses up through the aspens. During these rides, I kept whispering to myself a quote from Winston Churchill: "There is something about the outside of a horse that is good for the inside of a man." It seemed to be true, as the time riding was helping to relieve the mind chatter. I found myself taking every possible ride I could to have this escape, even if only for a couple hours.

Toward the end of the week, I got a call from the bank. A surprise since the company knew that I had taken the time to completely unplug. The call was chilling.

The deadline was now ten days away and I had no way to make the first payment. I had believed the advice that I would be OK if I missed the payment and that the bank was bluffing. What I did not know was that while I was gone, and without my knowing, Travis had casually let the bank know that I was not taking their threats seriously and that I thought they were just rattling sabers. My dismissing the "chain-the-doors" threat might have been appropriate in another situation, but Travis had provided the bank a viable path forward via his leadership that ensured it would get its money back even if they took control. The bank had believed in me and provided a path forward to prove it, and now I was thumbing my nose at its first deadline.

With Travis's help, the bank had tracked me down at the ranch to give me a succinct, stern message. "We understand you do not think we are serious. We are. If you do not have a $1M check in our offices in ten days, we are taking your company."

I put down the phone and felt like throwing up. I had to find a million dollars in ten days or my life would implode. The hourglass now had only ten days of sand left.

A MILLION DOLLARS IN TEN DAYS

OVER THE PAST MONTHS, I had become overwhelmed from the growing awareness that Starsys was now on a path to potential financial collapse. I was facing the reality that this had happened on my watch and that I was ultimately responsible for the situation we faced.

This truth had created psychological collateral damage in the form of a falsehood spawned from my growing self-doubt. It became impossible to shake the belief that since I had failed as a leader, I was incapable of leading us out of the predicament I had gotten us into. This crisis of confidence was quickly becoming a self-fulfilling prophecy.

Chris Hatfield, one of the more accomplished astronauts of the past decade, said it well: "Early success is a terrible teacher." The consistent, decade-long successes of Starsys were poor preparation for my guiding us through the situation we found ourselves in.

I was saturated with self-talk of, "We are here because of my failure of leadership," which made me blind to the parallel reality that Starsys remained a company capable of extraordinary results that could overcome a year of financial loss. After all, we were still a team of 150 of the best in the spacecraft business, uniquely capable of doing something of great importance for spacecraft makers, faster, and better than anyone else in the world. Seen this way and at the same time, considering the possibility of a cash infusion and a pulling back on the growth reins, it was conceivable that Starsys could continue to be wildly successful. However, this truth was shouted down by my recurring bouts of self-flagellation.

FROM THE GARAGE TO MARS

I felt like a pilot of a plane with a sputtering engine, frozen in the pilot seat, trying to come up with enough horsepower to miraculously get us over the ridge of mountains ahead to a safe landing beyond.

Because I saw us on a path that was going to take me, my company, and my family down, the sense of failure and indecision led to a hope-based strategy. I clung desperately to the belief that the next meeting with a potential buyer would reveal the white knight to save the company. Like Dorothy clicking her heels and declaring, "There is no place like home...." and waking from her nightmare of witches and malfeasant, apple-throwing trees.

Once the call came from the bank, I was forced to a crossroads. The easy path for me was to hang up the phone, wait ten days, ride the ship into the rocks, and then hope that I would be washed up on some shore, somehow finding a way to put my life back together on the far side of bankruptcy. A much harder path was to dust myself off, cowboy-up, and find a way through this challenge that seemed impossible to solve.

The call had been galvanizing. Like waking from being asleep at the wheel to discover your car is headed off a cliff with just enough time to grab the wheel and steer away from the brink. After the call with the bank, I felt myself casting off the shackles of my failure anxiety to look at Starsys with the optimism needed to survive.

I remember thinking about a quote from Franklin Roosevelt at that moment: "Courage is not the absence of fear, but rather the assessment that something else is more important than fear." I was still frightened, but I knew I needed to make objective, smart decisions independent of my fright. With the reality of what would happen in ten days now irrefutable, a different response to this situation was forced upon me and the need to take action was clear.

A MILLION DOLLARS IN TEN DAYS

I then made a call to our investment banker from a phone in the ranch kitchen, knowing that he would be a calming ally. In the course of our conversation, I was reminded that he had been developing a couple of investment opportunities with the diligence and focus that I lacked in my frightened state. Several paths had begun to form over the past month that might get us through. These would take time, but the odds were beginning to look like we might have a workable solution if we could keep the bank from acting too soon. It was possible to reframe the challenge to be "simply" keeping the bank at bay long enough to get to either a sale or investment that was developing. He smoothed my ragged countenance. He managed to ease my fears somewhat by pointing out that these opportunities were solid enough that we might be able to get individuals close to the company to participate in a bridge loan.[43]

As we continued our discussion, we talked through terms that might allow us to put loans in place quickly. He had confidence in the investment deals he was putting together, and he suggested that we could go to friends and family[44] with loan terms that could be viewed as a good, albeit risky investment. We began to believe that we might be able to raise the million dollars from a handful of individuals willing to put in a several hundred thousand dollars apiece.

As he outlined this plan, there was a growing confidence and shifting paradigm from "Why would someone want to invest in a failing company?" to "This could be a reasonably sound short-term

---

[43] A short-term loan used by entrepreneurial companies to bridge periods in which they are short on cash. The terms of the loan are such that the return is lucrative to the loaning party acknowledging the inherent risk in the loan. As an added safeguard, these loans typically convert to considerable stock in the company in the event it is unable to pay back the loan.

[44] When companies are formed, initial seed investment often comes from friends and family, or, in start-up lingo, F and F investors. This is sometimes broadened to F, F and F—friends, family, and fools—as these early investors are often not experienced enough in investing to properly judge the risk of a venture, and consequently are investing on faith in the founder and in the idea.

investment with risk but great return." Over the next 30 minutes, we worked through terms that seemed fair to both sides[45] and he agreed to write up the loan docs. My job was to reach out to friends and family to put together a million dollars in commitments in the next few days.

An unmistakable change had occurred within me during the half-hour, as I had gone from a nausea-induced panic to the thought that "maybe we can pull this off." As I hung up the phone, I felt increased confidence; "A bridge loan *was* good deal!"[46] I realized that I had simply needed someone to shine a light on this. I headed back to our cabin with cautious optimism warming inside me.

If I could get a significant initial investor, I knew that others would likely follow, a herd mentality creating a feeling of safety. And the individual I knew I needed to call first was Frank Tai. An early investor in the company and a ski buddy for the past couple decades, Frank was a friend who had stood up for Jackie and me at our wedding and was someone I sensed had enough liquid capital and appetite for risk to help us out.

I was fearful of the call, both because of the possibility that he would say no but also because I was asking for help, humbling myself in sharing how much trouble I was really in.

I decided that I would make the call on our trip back from Wilderness Trails to Boulder at the first point where I would have strong cell phone coverage. It was as if I needed someone to give me a countdown to brave a jump from a high dive, only in this case it was Verizon counting up: "One, two, three, four bars…Go!"

---

[45] The investment banker's recommendation was that each investor be guaranteed a 10% premium at the point of making the loan with 15% interest accruing for the length of the loan and the monies paid back (assuming monies were left over) even if the bank eventually took control of the company. This meant the lender could make $17,000 for loaning $100,000 for four months—good terms for both the investor and for Starsys.

[46] When you converge to the "right" investment terms (or in this case bridge loan terms), there is often a gut feel that manifests as "This is a such good deal, I should try to find the money to make the investment myself.". This is an important "tell" that leaks through to the investor during discussions and makes for an easier close. Once the banker and I developed the terms, the bridge loan no longer seemed so much as a favor to ask as a good opportunity being offered.

A MILLION DOLLARS IN TEN DAYS

As we came out of a mountain valley, halfway between Durango and Denver, I had the necessary coverage and pulled to the side of the road in the middle of not much of anything. I was in a desert, no cars for miles, mountains in the distance, with my family surrounding me in the car like a cocoon. Jackie was in the passenger seat, fully aware of the importance of the call. Ryan and Alyssa were watching a video in the back seat, oblivious to what was unfolding. Jackie and I said a short prayer together and I made the call.

After a few minutes of greeting and talking about the vacation, I dropped into the reason for the call with Frank, doing my best to squelch my fear and convey confidence. If he said no, the task ahead would again feel impossible. If he said yes, the path ahead would begin to clear. I was asking him to commit to a half-million-dollar loan. My anxiety was impossible to hide, but sensing this, he calmed me as I explained the situation. I came to the ask.

"Frank, I need your help. Would you be up for lending us $500,000 to get us over the hump and catalyze the rest of the investors? There's risk, but it's bounded. I'll likely need the money for four months until we find the right investor. The loan will net you about 70K in profit. What do you think?

There was a pause of a couple seconds, then a response that choked me up.

"I figured the call would come, but I'm surprised it took this long. Yeah, I'm in for $500K. Send me the paperwork and I'll look. If all is OK, I can get you the money in a couple of days."

The call had taken 10 minutes, his response 10 seconds. I ended the call with a weak, but cathartic "thank you" and everything changed in that moment. It wasn't that the risk was gone or that the path forward would be easy. Instead, I felt as if I had grabbed the stick of an airplane just before hitting a ridge of mountains, suddenly pulling up and sideways, and then turning to miss a looming mountain peak at the last second while simultaneously realizing that I could still fly the plane. There were still mountains ahead that were going to be a challenge to navigate through, but the play with the bridge loan had been a deft move that had worked. Getting help from Mike, taking the risk, calling Frank, and securing our first bridge loan investor re-instilled confidence that I could get us through this.

I made three more calls later that day—to my father and two uncles—all three of whom were willing to help given that Frank was already in for $500K. On Tuesday, I called the bank and stunned them with the news that I had $800K committed that I could deliver by Friday. With unexpected empathy, and now with a renewed sense of my commitment to a successful resolution, the bank said that the $800K would be enough as it felt I had met the intent of the first payment.

I had pulled off what had seemed to me impossible a week before. We had a plan moving forward, and we were working to a successful conclusion with the bank now firmly on our side. There was an extraordinary amount of challenge still ahead, but the hourglass had been turned over and we now had three months of sand left. My hand was back on the stick of the plane; I felt in control, still fearful, but again able to pilot.

SPACEDEV

COMING BACK TO BOULDER after the time in the mountains was like leaving a black-and-white tornado montage behind and opening a door to Technicolor. We were still far from home, but it was hard not to be a little bit optimistic when you see a yellow path at your feet that could possibly lead you out of the nightmare you were just in.

It *was* true the Starsys had value beyond the monies owed, and that it could be possible to garner substantial financial return over and above the debt if a plan toward this end could be presented without panic. It began to seem then that there were multiple paths forming, in what was shaping up to be a competition among multiple buyers, which might lead to a company sale at a competitive price.

A Deeper Look: Valuation

Valuation of companies seems like a dark art until you recognize that value is simply what a motivated buyer will pay, or what price a group of motivated buyers will bid it up to—the latter more accurately representing the fair market value. With a single buyer, the conversation is "Here is what I've decided your company is worth. Take it or leave it." With multiple buyers, it becomes a bidding war.

As one example of how this can happen, a friend and fellow Boulder entrepreneur who was not planning on selling his software company got a call from a major IT giant (for the sake of confidentiality, we will call the company Doodle). Doodle offered him 3X what he thought his company was worth for a quick sale. Over the next couple days, other Silicon Valley companies got wind, also called, and he was flown out to San Francisco several times for other offers. At the end, the resulting bidding war had him selling his company in six months to Doodle for eight times his initial personal valuation of his company.

FROM THE GARAGE TO MARS

Ironically, Travis became the catalyst. He was brilliant at explaining to potential buyers that the value of the company could remain quite high with a simple leadership change. In painting the picture of a founding entrepreneur/CEO struggling with growth, he explained how he could quickly turn the company around once he had the helm. It was a compelling story. "Starsys is a space company that has made a couple million-dollar stumble, but it can quickly right itself with the correct leadership and get back to a stellar growth track. I've got the experience to do that."

As Travis met with prospective buyers, I began to suspect that at some point he would intimate some version of "Scotty is a good guy and a great founder, but as you know, when a company gets to a certain size ..." His story played more strongly than if I had been pitching a turnaround solo: "I've learned some painful lessons ... we will succeed by limiting the type of programs we take on and throttle down the growth," which would have triggered a sense of skepticism rather than opportunity.

Although it was a bitter, bitter pill to swallow, I began to sense it was in my best interest to quietly align with Travis' story; to nod and go along with the premise that Travis would get us out of the quagmire. And in my addled state, it seemed not far from the truth. I was beaten down and new leadership looked to be the best path to a quick turnaround. At a personal level, the idea of relinquishing the helm with its associated worry and responsibility for 150 employees began to outweigh the unthinkable of giving up being the leader of the company I had created.

The response to my developing consideration of selling my company was "How could you possibly even consider giving up leadership of the company you created?" This is easier to wrap your head around when you realize the emotional toll arising from the level of worry that entrepreneurial leadership exacts. Whether times are good or bad, as the company grows a single individual is responsible for delivering on hundreds of promises:

- That investors who trusted you will be rewarded for this trust.

- That team members can count on employment and their payroll checks not bouncing.

- That your family will be financially rewarded for the risks that are forced on them as you work to realize your dreams.

- That you will attend your daughter's baseball games and your son's science fairs.

- That you will deliver on your promises to your customers.

- That the vision and mission of the company you have convinced your employees of will manifest.

- And that you will do all of the above with integrity and humanity.

As the CEO/founder, you must absorb all of this worry without passing it off on someone else (certainly not your family) while conveying with confidence that, "Of course, we are going to succeed greatly!"—despite the fact that at 3 a.m. the previous morning, you awoke with a shot of adrenaline and did not sleep for the next three hours while your noggin circled around some critical problem that was unsolvable in a half-sleep state.

All CEOs and entrepreneurs understand this. You can be sure that any of them will have an energized response when asked, "What do you do at 3 a.m. when you wake up with a problem you can't shake? When you've got to get back to sleep and can't?"

I found myself easing into this new thinking, "OK, Travis, you've got the helm. Now get us out of this mess." It was a strange co-dependency that was developing between us—me playing the part of the founder in over his head, he the cavalry, arriving in the nick of time. But it made for good story.

This arrangement came at great emotional cost. Travis walked into my office one day as we were preparing to meet with some investors. With a quiet intensity, he explained that he was willing to take the lead in pulling the company out of the muck, but only if I relegated my role as head of the company to him and appointed him president.

He then put a multi-page document on the table. As I read it, it became disturbingly clear that he and his attorney had spent weeks preparing for this ... waiting until my options going forward all relied on his involvement. My stomach fell. The document was thorough, covering things I had not considered. It stated that he would be appointed president, would have a seat on the board of directors, and that my role would shift to the relatively powerless chairman of the board.

The document also stated that Travis would receive a bonus of more than $100,000 if an investment were to be successfully closed. The document stipulated he would receive the full bonus amount as a severance payment if I relieved him of his duties after the agreement was executed. Implied but not stated was the fact he would quit if I did not sign the document. That in effect would indicate to the bank that I was not willing to do what was needed to save the company. If I were then to stumble, the bank would take control of the company shortly thereafter and Travis would be brought back to save it.

I could not help but be impressed with Travis' tactics. Although the company was being wrestled from my grip, Travis was showing himself to be a brilliant tactician. I discussed it with our board of directors, and they felt it was not far from extortion. They recommended that I tell him to walk and quickly find another C-level leader to help close the investment. My gut, however, was telling me that I needed to align with Travis' approach despite its egregiousness turning my stomach. I talked to a couple candidates, but I did not have the assurance they would succeed. As LBJ said, "Best to have your enemies inside your tent pissing out than on the outside pissing in." We were simply too deep in the river to switch horses and Travis was the only option. I had no choice but to accept the terms.

The only concession I was able to get was that Travis would assume the title of general manager; he would essentially be a second company president. Although the titles implied I was directing his role, it was clear that would not be the case. Travis would be leading the company. As major shareholder and chairman, I still retained full control of the investment decisions, including who we would choose as our investor and what terms we would accept, but I was unmistakably relegating leadership.

Although I felt I had been extorted, our partnership was strangely effective. Travis stepping into the leadership role to correct our course played well, the visionary founder in over his head, a new seasoned leader taking the wheel to steer us around the rocks.

As a result, several investment opportunities began to develop. One was from an investment banking group, which proposed paying off the loan, taking controlling interest in the company, and then bringing in a new management team (coordinated by Travis) to maximize return, followed by a sale of the company. With this option, though, there was no assurance that the original Starsys investors would see much, if any, return after we handed over control.

Although this was far from my vision of a transition and was a distasteful offer, it was a valid one. It was in effect a safety net under the high wire, giving us the confidence to explore other paths while calming our lending bank. And of course, the investment bank option was preferred to the bank chaining the doors. The offer started a process of us climbing out of the pit we were in. It was a platform that we could stand on, allowing us to scramble up to the next level.

Eventually a second opportunity came in: It was a bid from another small company, SEAKR Engineering, a Colorado space company that knew us well, that could see the value in our company. They were stoked about being in the space motor business, and that to my credit (and Travis' horror) were interested and willing to support me in the CEO position. On the call to discuss their offer, SEAKR led with a short preamble and then launched into its pitch: "We are prepared to immediately provide you one million dollars in cash." The offer was solid, but not as lucrative as I had expected. SEAKR would pay down our debt and provide a million dollars to the shareholders and myself. But it was less than we knew the company was worth.

After the call, Travis took me aside and earnestly conveyed that he was shocked, truly shocked. He believed that SEAKR's offer was insulting to me and did not deserve a response. I did not recognize that he was inoculating himself against a counteroffer that I might accept and would exclude him from the leadership role he was beginning to taste.

In my uncertainty, I agreed. I did not respond, time passed, and SEAKR retracted its offer, thinking we were no longer serious about selling. It took me months to realize that SEAKR had been prepared to counteroffer with a more generous proposal and was puzzled by my lack of response. By having me cut off communications with the company for the alleged insolence of its offer, Travis had created the intended result—eliminating the threat of a buyer who would have relegated him to a non-CEO role.

At that point, we had two offers on the table that I thought would sell us short, with nothing yet that reflected the value we saw in Starsys.

SPACEDEV

Then I was contacted by SpaceDev, a 30-person space company that had recently won the contract to build the rocket motors for SpaceShip1, a Paul Allen X Prize contest with a $10 million dollar prize to the company able to put the first citizen astronaut into space. SpaceDev was headed by Jim Benson, a visionary founder/CEO who was known as a bull prone to wandering into china shops. He was regarded in the industry as someone who could polarize the customers he was looking to court, such as NASA, by berating them for being less than competent. SpaceDev was developing a reputation as a company you wanted to steer clear of if you wanted to work with NASA. And that was the reason it was looking at us—we had the credibility, reputation, and gravitas needed to negate the loose-cannon reputation and take the company to the next level.

The previous year, Jim had realized he needed leadership help, and so he partnered with Mark Sirangelo, a C-suite exec cut from a different cloth than Jim. Mark saw great potential in SpaceDev, particularly if he could partner with other companies that might be purchased with SpaceDev stock. Starsys was the ideal target acquisition. It would give him almost 200 people instantly, and revenues far beyond those that SpaceDev was achieving.

When I heard that Jim Benson was interested in the company, I was less than thrilled. Most everyone I knew who had knowledge of SpaceDev said the company was not a credible buyer. That hesitancy changed when I met Mark Sirangelo. Mark was a compelling, provocative visionary. When you met him, you couldn't help but think, "He might just pull this off." He also radiated a charisma that was the polar opposite of Travis. I could see the company continuing under Mark's leadership in a way that I could get behind. The icing on the cake was that he had experienced a number of transitions of a founder/CEO into a merged company (including his own, as he sold one of his first companies). Despite the conventional wisdom that you should be cautious trusting the acquiring CEO, I had an inkling Mark was different, that he was the kind of person who was committed to successful founder transitions.*

In August, a couple days before my 30th high school reunion, I traveled to San Diego with Travis to meet the SpaceDev team and was impressed. It felt like Starsys ten years ago, a similar culture with a pinball machine in the break room, and 30 people working together in a way that seemed passionate and fun. I found my self-talk slowly shifting from "I'm going to lose my company" to "Maybe this could work." At the same time, there were two significant issues. 1) This would be a sale of the company, not an investment, and 2) my ownership and leadership of Starsys would be given up if we merged with SpaceDev.

---

*A Deeper Look
Founder Transitions: Handing Over the Baby

The transition for a CEO/founder from sole leader to working for the acquiring company is brutal and almost always a hot mess. "Only sell your company when you are ready to walk away from your company, because you will" is wise counsel. Every acquiring company will make the same pitch: "We love the culture and company you created. You are a key part of the changes we have planned to make this company a success. We are glad to have you on board. If you are not, you can leave and take your lucrative severance package with you."

This more often than not leads to a shouting match in the halls within six months, with the founder saying something along the lines of "Screw you! I'm out of here!" The founder then storms out, taking a cashed-out salary package while fuming about the company being stolen from him or her.

I can remember well a space company similar to us in size that was acquired by a company ten times larger. Up until the point of the transaction, the new owners had promised to "not change a thing" to preserve the unique company culture that had been in place for over 20 years. The day after the transaction, the new CEO walked in sporting a polo shirt and declared, "From now forward, this will be the official company dress code. In addition, we are removing the beer from the refrigerator, and we will no longer be doing A, B and C and will now be doing E, F and G." The resentment from the perceived bait-and-switch was substantial.

We captured talent as a result of that acquisition as the individuals who came to Starsys felt betrayed by the acquiring company.

SPACEDEV

More concerning was the nature of the transaction. It would be a sale for "paper." In trade for our company, we would receive stock in SpaceDev.

At face value, it could appear lucrative, but the stock price was artificially high, as it was based on a single contract that I sensed was tenuous. I asked Richard Slansky, the SpaceDev CFO, what would happen to the stock price if the contract were cancelled. There was a long pause as members of the SpaceDev team glanced at one another. Richard's hesitant response was, "That would not be a good thing. The stock price would tumble." The transparency was appreciated, as I noted that they did not sugar-coat but spoke the truth. But if we were to do a deal with SpaceDev, we would be gambling heavily on them pulling off what seemed to be a long shot.

SpaceDev was a public company, but its equity was considered a penny stock, which meant that the stock was thinly traded (few people bought or sold it on a given day). This meant the stock's value was somewhat artificial. If the public bought the stock, the price would go up. Conversely, if the public sold the stock, the price would fall, the public thinking that the buy or sell was indicative of something of import about to happen.

This is in contrast to a company on the New York Stock Exchange, where if you buy or sell $100,000 in shares of, for instance, Apple stock, it has no appreciable impact on the price. This same amount of buying or selling of SpaceDev stock would significantly impact the price of the stock up or down.

This situation creates a significant challenge for a founder selling a company in this way, known as "selling your company for paper." After the sale, the founder will naturally want to sell the company's stock to convert it to cash, both to take advantage of the newly acquired wealth, but also because of not having faith that the acquiring company will drive the price higher. But if the founder were to sell the stock shortly after the transaction, the stock price

> **A Deeper Look**
> **Protecting the Price**
>
> My intuition was correct. Within 18 months, the $40M contract was cancelled and the price of the stock fell to half of the initial value. In retrospect, I had missed a significant negotiation opportunity. I had been openly skeptical of merging with a small company without much of a track record.
>
> My arrogance was not intended as a negotiation tactic, but was accidentally brilliant, as their conversations after we met were guided by their thinking: "We have got to give Scott an offer he can't refuse." SpaceDev had been working to sell me on its vision and legitimacy, and I was not buying it. The company's reluctant admission that the price of its stock was fragile, coupled with its desire to woo us, created an opportunity for me to say "Then to be fair, the price of our company should take that into account. If your contract is cancelled, we should receive additional SpaceDev stock." But I did not. The stock price eventually fell, and the price we received for Starsys was halved.
>
> Once we agreed to merge, the final merger agreement did incorporate these types of terms to SpaceDev's favor. If Starsys did not perform as we projected, the stock promised from SpaceDev would be reduced. This also happened. We did not achieve all our goals. It was a $5M lesson learned.

would tumble with the public perceiving that the founder must believe something is wrong with the combined companies.

For this reason, the acquiring company will always require a lock-up period, during which the founder must not sell any shares (sometimes for years). Heartbreaking stories abound of entrepreneurs who sold Miketheir companies for a fortune, only to watch these same fortunes disappear without recourse as the stock price imploded. There are fewer stories out there of founders holding their stock for a couple years and then making ten times what they thought they would from a company sale.

* * * * *

We wrapped up our meetings. As I prepared to leave for the airport, Travis casually suggested that he change his flight to spend another day in San Diego.

Rather than having me take a cab, Mark offered to drive me to the airport, using the 30 minutes to connect CEO to CEO. During our drive, he suggested there were great things we could do together. As we pulled up to the Frontier Airlines entrance, he said, "Talk to you soon, my friend." He then shook my hand, and I hopped out into the sunny San Diego weather, thinking, "I could get used to this town."

I flew directly to Madison, Wisconsin for my high school reunion, and during the fight, I began to think that SpaceDev actually might be a viable option. I opened my laptop and continued my letter to my daughter, Alyssa, and then one to my son, Ryan, including the observation that I may have just visited the company that I might be merging with, and multiple trips to San Diego might be in my future.

When I arrived in Madison, I hopped in a rental car and drove downtown, walking into a group of a couple hundred former classmates at a downtown venue. I put my head down to delay being recognized, and then went straight to the bar, and ordered a Budweiser, thinking through how I was going to respond to dozens of "Scott—good to see you! Anything interesting going on in your life?" questions.

What I did not realize was that while I was in Madison having beers and catching up with classmates, Travis had returned to SpaceDev to continue the conversation. He also had recognized SpaceDev was a viable path forward for us, and he wanted time alone with Mark Sirangelo to explain to him the necessity of new leadership to ensure that Starsys succeeded greatly after the merger. Most critically, Travis would be ingratiating himself with Mark in hopes of becoming an embedded ally and a source of intelligence to ensure that the deal would close successfully.

DECISION

WE CAME TO REALIZE THAT SpaceDev was actually pursuing *us*. It gave us a badly needed boost of company self-worth; a Sally Field moment: "You like me! You really do like me!" SpaceDev was not simply sniffing around for an acquisition bargain. It had decided that we were the key to its growth strategy. It began conveying to us a vision that together, we could do great things beyond what Starsys could do alone. It wanted to find a way to merge, retire our debt, and give us the resources to recover from our rough year.

Starsys being considered a cherry of an acquisition target changed everything for us. With the SpaceDev opportunity legitimate and developing, we began to once again see Starsys as a prized company despite the previous difficult year. Where others had seen us as distressed[47] and warranting only a fire-sale price, SpaceDev was looking at us as an acquisition opportunity that would allow it to *Go Big*.

Although this promising turn of events gave me the needed confidence to frame Starsys in a light so that others could also see the value, I was not convinced that SpaceDev was the right partner. Because the company's offer did not involve cash, the proposed transaction would provide mostly SpaceDev stock as payment. It looked good on paper, but the stock would not be tradable for years,

---

[47] "Distressed company" is a term in the world of mergers and acquisitions that describes a company in financial trouble, and as a result can be purchased for far less than it could be if healthy. If the cause of the distress can be remedied (for instance, by an infusion of capital), the value of the company quickly increases. Similar to *Flip This House*, it is like buying a fixer-upper home at a bargain, righting the few things that are wrong, and turning a quick profit.

and as a result Starsys' value would be reliant on future SpaceDev successes. A far from sure thing.

My reluctance was palpable and worked in our favor, as SpaceDev sensed it would need to make an offer we couldn't refuse.

During this time, I had been hamstrung within the narrative of "Nobody will give us the $6M we need to pay off the loan without taking control of the company, therefore we've got to sell the company lock, stock, and barrel."

Then another option began to form. Since Starsys was distressed, its value was depressed. Would it be possible to create a syndicate of sorts to provide a $6M cash infusion, giving the syndicate less than half the company while remaining in control, correcting what wasn't working, and then selling the company for a value that might be more than $20M? The math looked reasonable. The syndicate would get a substantial part of the company for $6M, and then we would sell in a couple years for more than twice that.

As it turned out, being able to say "we have a highly motivated buyer in SpaceDev, but we are not sure they are the best partner for us" was the tipping point. It provided the credibility needed to assemble a group of investors to make a competing offer that would retain our control and ownership of the company.[48] The team we put together was a bit like the overused movie trope a ragtag group of individuals, a la *The Magnificent Seven, Oceans 11,* or *The Breakfast Club*, disparate in their backgrounds, resources, and motivations, who come together to do something of great import.

In this case, the group included one of our Japanese customers wanting to invest in the US, an eye surgeon/angel investor, the brother-owners of SEAKR engineering who continued to want to

---

[48] This is similar to what happens in public companies and that often makes the news. When a takeover bid is being put together to buy control of a company and the current owners of the company are looking to retain control, the existing owners create a consortium of lenders, investors and cooperative partners to provide a competitive bid that hopefully prevents the takeover.

DECISION

invest in Starsys, another space entrepreneur with extra cash on hand, and a couple of other informal investors. With a bit of bailing wire and duct tape, I was confident we could put together the $6M that the bank needed to pay off the loan.

The plan was that we would say "no" to SpaceDev, gather the $6M, pay off the bank loan, quickly turn Starsys around to profitability, grow the company further, and provide a two- to three-time return to each of the investors.

My share of Starsys would be diluted from 50 to 25 percent, but I would remain as CEO and in control of the company. I wouldn't have to give up my baby.

I tried to ignore that in essence, I was building a house of cards, where each of the elements relied on the others to succeed. If any one of the partners stumbled, it would fail; but I was only willing to see this cup as half full. I wanted to keep my company.

In early August of 2005, I headed down to Denver to our investment bank to meet with one of the investors, Eric Anderson, who would be responsible for the majority of the syndicate investment. If we could come to an agreement after walking through the investment terms, it would provide the green light to bring the other investors on board and I would have a path forward to paying back our bank and keeping control of our company.

Unbeknownst to me, Travis had kept a close eye on the developing syndicate deal and had continued to apprise Mark Sirangelo of my growing enthusiasm for a parallel path. Based on what was to happen later in the day, Travis had given me a gift ... a call to Mark as I left for Denver. As I was to later infer, he had told Mark that I was leaving to meet with Eric Anderson in Denver and that I was pulling the investment details together with the rest of the investors. He also told Mark that it would be about an hour until the meeting started and that he would have to make me an offer that would hit it out of the park.

FROM THE GARAGE TO MARS

Backchannel communication like this, with Travis keeping SpaceDev aware of our competing opportunities, would have been a masterstroke of negotiating strategy if I had had any clue that it was occurring, but I was blissfully oblivious to the G2 [49] being shared with SpaceDev. It turned out to be a blessing in disguise, though, as I would have been sure to muck things up had I had known about the communication between Travis and Mark Sirangelo.

I pulled into the parking garage and rode the elevator to the investment bank's offices, equipped with the requisite large boardroom table, leather chairs encouraging insouciant slouching, and a plate-glass wall overlooking the 16th Street Mall in downtown Denver.

Eric arrived shortly after I took a seat. It was the first time we had met in several months. Once some initial pleasantries were exchanged, Eric made it clear there was only one last issue on the table before all would invest. The group of investors wanted its portion of the company to be close to 50 percent at the end of the investment, so that if something did go south, and additional monies were to be provided by the group, it would simultaneously gain control of the company. This was something I could give on. It was reasonable and with my optimism that the company's problems had been fixed, it was something I was prepared to agree on while I worked to not appear too willing.

As Eric and I talked, the fax machine in the other room began to spit paper. A few moments later, one of the partners knocked on the conference room door with a fax in hand and a face trying to look not excited, and failing.

"Scott, sorry to bother, but I need you to take a quick look at this."

---

[49] An oft-used term in the space and military industry (and associated businesses) for "intelligence information regarding an opportunity or opponent." In WW2, G2 was the organization within a general's command that provided military intelligence; the phrase "any G2?" basically means the following: "Is there any intelligence on this that we were able to obtain by sneaky means?"

DECISION

I stepped out and he smoothed the spooling fax paper on a nearby desk and walked me through the terms. The offer was from SpaceDev and it was close to twice as good as what the company had provided previously and four times better than any other offer from any other company. The proposal was lucrative at the time of sale, but made even more mouth-watering with SpaceDev's inclusion of an "earn-out." If we achieved the results we had projected over the coming couple of years, the monies we would receive for the deal could possibly triple. All we had to do was deliver on what we said we would do.

In addition, the fax had an element that I had not really put much thought into. It included a very sweet deal for myself as the CEO that would increase my salary over a period of three years, with an implication that my salary would continue whether I ended up working for the company or not. I was trying to look at this deal solely as a fiduciary of the shareholders, but I could not help but start thinking about what a three-year paid vacation might be like.

But that inside voice was whispering again, "You would be giving up your company!" On top of this, the deal was primarily structured to be based on a transfer of stock. If the value of the stock dropped, the lucrative nature of the deal evaporated with the falling stock price.

However, if Mark Sirangelo was able to do what he said with the company, the value of the stock could double, or triple and Jackie and I would end up wealthy and able to retire before I was 50.

But, I would be giving up my company.

As I walked back into the boardroom, the investment banker's partner made a final comment: "I don't think I've ever seen a deal this lucrative in this kind of situation. How can you pass it up?"

I reentered the room with an awareness of the deal in the back of my head, but I almost wished I hadn't known. I really wanted the deal with Eric to work.

We talked a bit longer about some of the details. The ideas that Eric was presenting would allow me to move Travis out of the company and retake the helm. The SpaceDev deal became less interesting minute by minute as I envisioned our company being back on track with me in charge.

At the end of the meeting, I had convinced myself. I stood and said, "Let's do this!" and reached my hand across the table to seal the deal. I knew the handshake wasn't binding, and I could still back out. However, a part of me wanted that extra level of commitment. As I digested the SpaceDev offer, I knew it was going to be hard to say "no" once I fully understood what was being offered in recompense for my giving up my company.

But something was gnawing at me. The consortium deal meant I would retain the personal guarantee on the loan. If we were unable to turn the company around in a couple of months and were to continue to lose money, I would be back where I was, having to peddle a still-distressed company with an additional loss of value and leverage that would surely result in the loss of the company, my home, and my family.

I was sure we had learned our lessons. We weren't going to make the same mistakes again, and we would quickly be back on track and profitable. Right?

I was stuck and could not decide, and time was passing. There were two legitimate offers and I had to pick.

If I were to pick SpaceDev, I would be handing my baby over to another parent with an uncertain reward. It was so troubling that I had dreams of handing over a small model of our building (with "Starsys" on a sign on the side) to the new owners. In this dream, I

DECISION

couldn't let go of the building; they had to wrestle it from my arms. I woke up frightened and weeping.

If I were to pick the syndicate, I would be back in the captain's chair, but it felt like I was fooling myself. There was no certainty that we had turned the corner to profitability. I tried to ignore the fact that there remained significant operational issues that hadn't yet been resolved. With entrepreneurial optimism, it was not difficult to convince myself and the rest of the consortium that we had it all worked out. But with pragmatic realism, I couldn't shake the idea that the consortium deal would have us back into the maelstrom within a couple months, with a boat that couldn't withstand the storm.

I remember standing in my walk-in closet, stuck thinking that this is one of those leadership moments where the leader commits to path A or B, without knowing for sure which is the right path, but having to make the decision, nevertheless. In that moment, I tentatively decided that it was time to man up, do the syndicate deal, say "no" to SpaceDev, and get back into the saddle.

What was missing was the calm that I expected from committing to a particular path. The decision had been made, but I was far from at peace. I did not feel like I was following a Nudge. I felt instead that I had made a decision based on what I was afraid I was losing, rather than one based on what I would gain.

I hesitated to let SpaceDev know that I had chosen not to accept its offer. I knew that making the phone call to Mark Sirangelo would be closing a door forever.

I was out of my mind with indecision. I could not make the call. The Nudge that I'd come to trust was not goading me forward to the consortium—it was whispering "Let it go." I didn't want to hear that. I was stuck.

It was August 22nd, 2005. My wife's birthday. We had finished dinner and I was sitting on the front porch. It was a hot summer

> **Tibbitts Tip**
> **Let Go or Be Dragged**
>
> Of the life and business lessons learned from the Starsys journey, this is one of the most powerful for me, and for life in general. We cling to what we know. We hang on far too long. We don't want to let go. Yet, we cannot move on to the important "next thing" for us.
>
> In my latter venture, Katasi, I wrote this as a mantra I kept on my desk as I weathered storms that threatened to capsize the ship, so that I was cognizant that "the captain must go down with the ship" is unwise counsel.

night, and I remember bathing in the yellow light and the warmth and sounds of the evening with the mosquitoes circling the can lights in the porch ceiling.

I had called Merc Mercure, an entrepreneur, CEO, and a mentor of mine, who had a storied and legendary reputation in Colorado. I was very fortunate to have this sage, whom I could call and ask for council, on my side.

I explained the situation. I expected Merc to back my decision to find a way to make it work; to hunker down, take the consortium's investment, and make it so.

Instead, he paused, reflecting. Quite a bit of time passed before he spoke, as he must have realized I would take what he said to heart, and that his advice would dramatically affect my future and Starsys' as well.

"Scott, that personal guarantee will kill you, literally. If anything unexpected happens, the bank will come after you and take everything you own. You will lose your home, and possibly your family. Scott, sell your company. Take the reward you've earned. Life is short. Starsys has had a good, long run on your watch. It's time to move on. Let SpaceDev take the yoke and pull."

I remember his words, the warmth of the night, the mosquitoes circling the lights, as if it were yesterday. And then I felt the goosebumps. The Nudge.

"Thanks, Merc," I spoke. "I get it. Thank you. Talk to you soon."

DECISION

I stood up and walked into the house, finally feeling the peace I had been seeking for more than a year. The next morning, I called Mark and said, "Let's do this thing."

I hung up the phone, knowing that I had just committed to giving up the company I had invested 20 years of my life to create.

FOUNDER IN AN ACQUIRED COMPANY

I THOUGHT IT WAS A DONE DEAL when I said "Yes" to the SpaceDev merger. That the rest would be just details. After all, I had agreed to and executed a terms sheet that was several pages long, including all the key elements of the transaction. Our investment banker called shortly after the terms sheet was signed, and, while I was in a "woo-hoo!" mood, threw water on my celebration.

"Slow down, Scott. Yes, great news, but let's keep it to a mini high five for now. We have a long way to go before the deal closes."

I soon understood what he meant. There were hundreds of details to incorporate into the documents required for execution, and, in parallel, there was the need to support an excruciating process of due diligence to enable SpaceDev to confirm, as is the norm for an acquiring company, that all the things that we had been saying about Starsys were true and there were not any skeletons in our closets. Each of hundreds of details with our merger needed to be negotiated and legally codified. Multiple times during the process, I would push for something, and Mark would respond, "Scott, that is a deal breaker. Can't do it."

Mark was an extraordinary negotiator. When he suggested that a particular issue was a walk-away point for SpaceDev, I would usually capitulate, because if the deal came apart now, there was no going back. We were actually so fully committed at this point to merging with SpaceDev that the bank would take the company if the negotiations faltered. Along with our investment bankers, we were collectively holding our breath, as it would not go well for any of us if this were to unravel. What I didn't realize was how much Mark and SpaceDev were in the same boat ... he had committed a million dollars to the bank that was not refundable, and he did not have a plan B. "We have to get this done no matter what it takes" worked in both our favors and led to what we both considered years later to be a fair deal.

A key element of the transaction negotiations was the earn-out. We had assured SpaceDev that we would be quickly profitable with its infusion of cash. SpaceDev in return leveraged this optimism by

---

*A Deeper Look: Earn-Out

There are a couple of terms that serial entrepreneurs learn to be very cautious of and the "earn-out" is one of them. When you are selling your company, you typically project a rosy future to the acquirer to help convey the value of the company. A seasoned purchasing company responds, "Gosh...you are worth quite a bit when you make these forecasts ... how about we pay you a portion of the value now, and the balance after you've met the goals you are confident of?"

There are three factors that conspire for this not to work out for the entrepreneur:

1) The new leader may not be as good as running your company as you are (which is often the case) as you gave up the control necessary for keeping the company on course.

2) The irrepressible optimism of the entrepreneur can't help but leak into the projections (the numbers are something the entrepreneur is convinced will manifest but rarely do). And ...

3) Once an earn-out is in place, it is absolutely, positively, almost always in the best interest of the new owners of the company to work the financials to miss the goals until after the earn-out period is complete.

Earn-outs are notorious for not paying out. The forecast presented by the seller is optimistic so as to maximize the valuation. The acquirer knows that the goals are unlikely to be met and uses that advantageously. In addition, if the earn-out is based on profit (as it was with Starsys), the acquirer can adjust the finances to purposefully miss the profit goals. To inoculate against this, a seasoned entrepreneur will lessen the goals to achievable and base the earn-out on revenue, not profit.

FOUNDER IN AN ACQUIRED COMPANY

proposing a deal in which the price it paid for our company would double if we met the promised numbers. As described in an earlier chapter, an earn-out is a seductive siren call that preys on entrepreneurial optimism. This was made worse by the earn-out being based on profit (how much money we made) rather than simply our revenues (how much product we sold).

Profit-based earn-outs immediately create cross-purposes between the buyer and seller. In this instance, SpaceDev was rewarded for keeping profits low, while we as the seller (Starsys) looked on helplessly, knowing the company could be more profitable but having no ability to influence the outcome short of legal action. As it turned out, SpaceDev did not engage in any overt malfeasant monkey business, but at the same time, it was certainly not in its best interest to maximize profit during the two-year earn-out period, which gave little chance of realizing an earn-out.

And then there was the due diligence. Travis' involvement in the company had been both a blessing and a curse, and in this case, now that the transaction was locked down—a transaction that he was stoked to see coming to life, he leaned in and pulled hard to make the merger a successful, smooth process. Travis had the team invest weeks in creating stacks of three-ring binders that were shipped to SpaceDev and included every possible company detail.

---

*A Deeper Look: Due Diligence

An offer is made to buy a company based on the information the seller (in this case, Starsys) provides during the sales phase. It was in our best interest to present our company in the most positive light possible, but without exaggeration, as the promises made would be validated during a multi-month due diligence process in which the buying company (SpaceDev) would review every detail of Starsys, warts and all. This is analogous to doing a home inspection after being under contract. If anything unusual is discovered, or any misrepresentation found, the terms can be renegotiated or the buyer can walk from the deal.

FROM THE GARAGE TO MARS

The strategy was smart. The data was extensive and filled with a multitude of immaterial facts. If there was any dirty laundry in the company, it would be almost indiscoverable within the reams of data provided. Finding a fly in Starsys' ointment would be like finding a needle in a haystack. We just sent them pretty much every record we had. We were not aware of any "gotchas" in the records; but if there was something amiss, something that could have affected the deal, it would be deep in the mountain of documents we had provided and would be hard to find.

August passed, and the deal close started stretching from October, to November, to December, and then into 2006. Everything finally began coming together for a January 31st, 2006, close. By early January, it was clear there was nothing that could prevent the transaction from going through, and I began to relax.

Then, two weeks before the closing, I was informed that an audit of our export approval process uncovered a problem of significant consequence.

Spacecraft hardware that is exported outside of the country and that helps a non-US country build spacecraft can be construed to be enabling the importing country's military capability, as many defense systems are space- or missile-based. Because of the potential threat posed by these defense systems, this type of trade is controlled by the International Trade Arms Regulations (ITAR), which constitute an extraordinarily difficult and complicated series of hoops and hurdles to work through to get the OK to ship hardware outside of the US. At Starsys, we worked hard to operate within these rules, as one of our international programs provided launch vehicle separation systems for Sweden. We had been complying for years with ITAR and had painstakingly followed what we thought was the proper protocol.

In the final stages of due diligence, it was discovered that we had been following the wrong protocol in shipping more than 30

FOUNDER IN AN ACQUIRED COMPANY

> ***A Deeper Look: Selling for Paper**
>
> If a company is publicly traded, a portion of the payment is often made in stock ("selling for paper") and the previous owners are not allowed to trade that stock for a couple of years. There is a significant chance the stock will go down, reducing the return. In the case of Starsys, we lost 50% of the value of the transaction because of a dropping stock price.

devices over two years, out of the country, without the proper US State Department approvals. In the past, large companies within the US that violated these protocols have been fined tens of millions of dollars. In the same way that finding a house riddled with termites during inspection kills the deal, our export violation was exactly the kind of due diligence discovery that could kill the transaction, as SpaceDev would be on the hook for any US State Department fines once the merger transaction was completed.

As a result, I went crazy with panic. Mark Sirangelo was also becoming gravely concerned, leading to a tense "Scott, how is it possible something like this could have happened?" conversation. Our saving grace was that he was also all-in on the deal, with much to lose if he had to pull the plug, as he had convinced his board that Starsys was flaw-free and worthy of investment.

Adding salt to the wound with the twisted enthusiasm of a boy pulling the wings off of flies, Travis popped his head into my office shortly after to casually let me know that CEOs have been known to go to jail for this type of thing. His words only served to pour additional gasoline on my panic-fueled inner bonfire.

We took a figurative deep breath and attempted some mitigating measures, which included hiring a lawyer to talk to the State Department. Our lawyer explained the situation, stating that it was a minor administrative blunder rather than an intended end-run around the regulations. We began to hear that this was likely to be considered an honest mistake, one that if corrected would be forgiven. We immediately began the process of redoing the export license paperwork. Fortunately, Mark was well connected to the State Department, and he also began to hear similar feedback.

To go forward, he would have to convince the SpaceDev board that completing the transaction was worth a potential multi-million dollar fine from the State Department.

It came down to a final board meeting to approve the deal or not on the afternoon of January 31st. It was out of my control. There was nothing to do but wait for the board's decision that would determine if Jackie and I would be ruined financially, or if we would move forward and Starsys would be able to make the transition successfully to its next chapter.

The meeting was to start at 1 p.m. and last one hour. I sat at my computer waiting for the call. Not much to do but watch the phone. I could feel the tension within the company. We had told our people it was a done deal; we were just waiting for the formality of the OK, but they knew differently. Our transparent culture meant that every one of our 150 employees knew about the export issue and what hung in the balance. Every five or ten minutes, someone would pop into my office with a forced-casual "Any news?" All I could do was offer my equally forced-casual, "Nope, not yet. Any minute" response.

Two hours passed, then three. It was horrible. The approval process was supposed to be short. However; the more time passed, the less chance there seemed to be of success. I was becoming sure that the reason for the lengthy board meeting was because everyone was talking through the details of backing out of the deal.

And then my CFO put his head in the door and said that he had been asked by SpaceDev for our bank's wire transfer information. There were a few minutes of confusion as the certainty of the deal failing seemed to be swirling with what felt like a whisper of very good news.

FOUNDER IN AN ACQUIRED COMPANY

The phone rang a few minutes later. It was Mark. The deal had been approved, and the documents had been signed. It was done. He congratulated me on now being a part of the SpaceDev team. I could hear the strains of the Kool and the Gangs' song "Celebration" playing in the background over the intercom at SpaceDev HQ. As relief swept across my body like a wave washing me clean, I said, "Thanks, Mark. We're going to do great things together!"

I hung up the phone, sat for a while quietly with the news that no one yet had, and soaked in the feeling that we had actually pulled it off. We had made it through hell. One door had slammed shut and another had opened—all in the period of a 60-second phone call. I gathered the company to let them know, and then headed home to tell the family.

I slept a full eight hours that night, the first full night's sleep I had had in nine months. I woke late to a brisk, bright February 1st, 2006, morning, as a founder in an acquired company.

THE RED BUTTON

It was surreal driving to work on February 1st, an un-responsible leader. There was no longer a substantive, irreplaceable need for me in the company I had created. I could have driven past the company that morning, headed to Denver International Airport, hopped on a plane to wherever for a month of depressurization, and eyes would not have batted. It was both stress-relieving and self-worth crushing.

The relief of no longer being responsible had unintended consequences. Unbeknownst to me, my panicked amygdala had been screaming fight or flight throughout the ordeal, marinating me in adrenaline for nine months. It took a toll. My adrenal gland was exhausted and shut down like an Ironman competitor collapsing on crossing the finish line. It had had enough. With adrenaline production on pause, the adrenals similarly stopped producing cortisol, my body's joint lubricant and anti-inflammatory agent.

Every joint in me began to ache from the absence of the hormone. It was a struggle to get out of a bathtub or put on a suit jacket. I was moving like an arthritic 90-year- old. After a couple of weeks of "What the heck is going on?" I went to an endocrinologist to find I had polymyalgia rheumatica (PMR), a disease that typically affects women over the age of 75.

Lovely. The stress had given me an old-woman disease.

The protocol for addressing PMR required me to overwhelm the lack of cortisol (SR) with a synthetic hormone, prednisone. A day after taking the medication, the aching was gone, and my joints felt miraculously lubricated; but the side effects were brutal: My face swelled like a puffer fish from water retention and the hormones ran roughshod with my emotions. The slightest comment would trigger a cascade of disproportional responses. "Hon, these eggs are a bit overdone..." from my wife elicited "So you're saying I'm a failure as a father and husband?" I gained empathy and appreciation for PMS, as my endocrinologist explained that prednisone created a similar hormonal emotional roller-coaster ride. The PMR protocol required the prednisone to be taken for a year with a tapering of the dosage until it was no longer needed. It was a long nine months until my endocrine system rebooted and I no longer needed the hormone.

Although my responsibilities were limited, I was being paid well to continue to be the face of Starsys. It was important to SpaceDev that the transition from my leadership to Mark Sirangelo's be as seamless as possible, with my continued presence signaling an amicable hand-off of control. Mark was a blessing to the process. He had been involved in multiple CEO transitions post-acquisition, being both the CEO transitioned from and to. He understood the emotional elements at play and did well in gentling me as well as possible through the process.

I retained my office and was given a new title, "Managing Director," which was generous given there was little I was still managing. It was a strange situation, to say the least, to still be involved in a company I was no longer leading.

The experience of being a founder in a merged company is conveyed best by thinking of it as you would an amicable divorce. You've separated from the family you've loved and grown up with. Your wife has remarried. But despite the divorce, you are given a beautiful room in the house in which you grew your family, with comfy furniture, a thick luxurious carpet, and a picture window that lets you watch the new husband raising your kids. You are allowed to provide advice, but without authority.

THE RED BUTTON

One morning, you are looking through the window, and to your horror, you see the new dad feed Twinkies to your kids for breakfast. You leave your room, go into the house, and say, "Hey, Tom, I'm really not sure that it's the best thing to give my kids. Wouldn't yogurt and berries be a better choice?"

And Tom says, "Yeah, thanks for the input, Scott, but no, we're not going to do that ... Hey, kids, want a couple more Twinkies?"

And so, you shuffle back to your nice room, emasculated, waiting for the next "that's not how it should be done" moment that you can do nothing about.

To my surprise, I stayed for three years. I had been through the wringer and needed that time to lick wounds and heal. Mark knew this and gave me a safe harbor to sort out what was next for me. The "f-you, well f-you" conversation that often happens between the founder and the new CEO never happened. Mark worked with me to have the transition be as smooth as possible to set SpaceDev up for success. Although I was no longer at the helm, these were 150 people whom I cared deeply about, and a mission that I had started and wanted to see Mark continue.

To paraphrase Dickens, these were the worst three years of my career; they were the best three years of my career.

Watching someone else raising your family, you come to understand the things that are your strengths. I could see the things I had brought to the company that were missing with the new leadership. Competencies that were natural, a part of my being, that I was surprised to see came hard for others.

More importantly, I came to recognize what I had not been good at. Leadership elements I had struggled with. Things I could see in retrospect that, frankly, I had sucked at. The ruts I had created. The dysfunctional patterns we had been stuck in, that at the

FROM THE GARAGE TO MARS

> **Tibbitts Tip**
> **Stop Doing What You Suck At**
>
> A precious leadership (and life) lesson I learned the hard way through the Starsys journey was to realize we are all blessed with things we are great at wired into our DNA. Things that showed up early in our lives that come naturally to us, but are mystifying to others who struggle to do the same. Our brilliances.
>
> Similarly, we all have shortcomings that we struggle with, that are unnatural for us. The things we suck at.
>
> In my case, simply stated, I suck at operations and finance. I'm brilliant at vision and leadership.
>
> Our ability to lead is transformed when we recognize and embrace our brilliances, as well as our limitations. And recognize there are those who love what we struggle with and who are challenged by what is easy for us.
>
> Find that partner. Hand off those things you struggle with and focus on what you excel in. Follow that which brings you joy.
>
> If you struggle with knowing what you struggle with and where your brilliance lies, simply ask those around you. They have been aware of it for years.

time looked to simply be bad luck, but instead, started to look inevitable given the way I had been leading.

I had been skilled at creating and conveying vision. In creating followership, loyalty, and company buy-in. In creating cool products that solved big problems. In creating an organizational culture that attracted the very best talent and customers. My awareness of these strengths was salve on my wounds as I reconstructed my self-worth.

At the same time, I could now see my Achilles' heels. I had been unable to make hard calls. To quickly fire those who didn't perform. To tell our customers the occasional "No" rather than a constant "Hell, yes!!" To say, "I'm sorry, but that is what it costs. Take it or leave it."

My leadership had created a company that everyone wanted to work for and with, but left it with a soft underbelly, vulnerable to failing operationally when things got tough. Combining my desire to not disappoint the customer with an insatiable desire to take it all on and grow had set us on a course upon which it became inevitable we would founder.

THE RED BUTTON

As I talked to other entrepreneurs about their challenges to growth, I came to recognize entrepreneurial leaders are wired with unusually strong suits that allow them to do what they do: to carve something of substance from nothing but an idea. However, the boldness that engenders "if you can dream it, you can do it" thinking comes with a blind spot: "I can do anything if I put my mind to it" becomes "I'm great at everything I put my mind to."

The optimism and blue-sky, we-can-do-anything thinking that drives the ability to create something from nothing is the antithesis of the caution and risk aversion necessary for effective operations as a company grows. The empathy and rapport-building skills necessary to create deep, trusting relationships with investors, customers, and employees is in opposition to the hard-line pragmatism required when the tough calls need to be made that can cause pain and anger and end friendships. The promoter personality who can paint a Technicolor picture of the future is poorly suited for the day-to-day planning minutia necessary for that vision to materialize.

While serving on a panel several years later with several CEOs who had been invited to talk to business students aspiring to be entrepreneurs, the following question came from the audience: "What are you NOT good at as a leader?" The enthusiasm among the panelists to answer that question was remarkable. Hands shot up. "I am NOT good at operations." "I suck at accounting and finance." "I'm a terrible internal CEO ... I need a general manager to run the company while I build customer relationships." The level of ownership (and enthusiasm) of what we all were deficient in was eye-opening to all in the room.

> *I came to understand one of the more powerful attributes of a leader is not simply to build on what you are brilliant at; that is something that the self-aware quickly grasp through feedback from those around*

them. The harder skill is assessing and owning the areas of your leadership deficiencies, the short suits, what you are terrible at. And then finding leadership peer partners who are brilliant at your deficiencies. The leadership yin to your yang.

My "aha" was not that I should have gone against my natural grain and learned to say "no" to a customer as much as I had needed a partner wired to say "take it or leave it" with integrity. I needed to give that person authority, and free myself up to spend time in my strengths: setting vision; getting great talent on board; developing close, trusting relationships with customers; and pointing at a mountaintop and saying, "Come on, we can do it. Let's go there!"

You don't need to look far to see the necessity and power in such peer partnerships: Steve Jobs and Tim Cook at Apple. Bill Gates and Steve Ballmer at Microsoft. Larry Page and Eric Schmidt at Google. Each a partnership between a visionary and an operator that became legendary.

* * * * *

Mark Sirangelo had a big dream for the merged companies. The space shuttle program was winding down, and NASA was looking to commercial space companies to build a replacement. Elon Musk and SpaceX proposed we return to Apollo-era capsules to get astronauts back and forth to space. Safe, but far from sexy.

NASA wanted a competing technology that would be forward-looking rather than a return to the past. A spaceship that flew, not one that simply cannonballed out of the sky under a parachute. Mark's vision was to use the gravitas of the combined companies to propose a smaller winged spacecraft, called Dreamchaser, sized for one role: to take six astronauts back and forth to space. It was an audacious dream, the idea that a commingled company with only

THE RED BUTTON

Dreamchaser
(Back row: the author second from left, Mark Sirangelo far right)

200 employees might be trusted by NASA to build a space shuttle replacement.

It took more than five years and three tries, but in 2016, Mark pulled it off. The company that a bunch of ragtag, wannabe rocket scientists had started and that later became SpaceDev (and then renamed Sierra Nevada Space Systems after a subsequent acquisition in 2008) was awarded a multi-billion contract to build the Dreamchaser for NASA to resupply the space station. It is scheduled for its first orbital flight in 2025.

It is difficult for me to wrap my head around the idea that an invention made of copper tubing and wax begat a spaceship that will provide transportation to and from the Space Station. But if we hadn't put wax in that tube, poured hot water on it, watched it extend and said, "Whoa, that's cool! I wonder if NASA could use this thing?" those 17 people pictured opposite would not have been standing in front of a spacecraft 20 years later.

Hearing of the Dreamchaser award from NASA was an extraordinary moment for all of us, but it is not the achievement I'm most proud of.

FROM THE GARAGE TO MARS

That is reserved for the group of engineers, technicians, and former burger-flippers, most of whom had never dreamed of working on spacecraft, who had all lived the mantra "It absolutely, no kidding, has to work in space, no matter what it takes." All had fun and worked as a family, living fulfilling lives as we made rocket parts. They were people we trained to be rocket geeks; people we flew on the vomit comet with; people we celebrated marriages and additions to families with; and people we cried with through personal tragedies such as Kurt's passing and national tragedies such as 9/11.

The Starsys family we had created over 20 years built more than 4,000 mechanisms that flew on more than 350 spacecraft with zero failures—a record unequalled by any other space company in the industry.

* * * * *

I've imagined being able to go back in time and talk to my 2003 self from the perspective of what I know now. I envision it as a lucid dream; my past-self sitting with future-me, high in the Wyoming mountains, perched on talus near a small lake surrounded by peaks.

In the dream, the past-me knows that what is being heard is truth. Counsel from my future. What would I say if I had only a few sentences that I could convey, but would be believed and acted upon? I wouldn't need more than 60 seconds.

- "Say NO to programs that would drive our company to grow faster than 20 percent per year."
- "If you take on big programs that become troubled, stand firm for more money, or walk away from the program."
- "If a truly talented leader offers to partner with you to make Starsys great, say "YES!" despite any reservations you may have.

THE RED BUTTON

Any one of those lessons heeded would have saved millions of dollars, and I likely would still be leading Starsys as a rock-star, rocket-scientist CEO.

Would I change things if I could? If, in some magical way, I was presented with a big red button and could press it and change the past, would I do so? I can imagine it next to me, on the patio table, in the shade of the umbrella alongside the pool, where I chose to write from this morning. With a cool breeze blowing, the button teases me with the promise of an alternate past-future.

Everything would change. The life lessons learned from a corporate near-death experience would be lost. The opportunities and the amazing "what's next's" that opened following the door closing on Starsys: a day with Katie Couric driving in Manhattan talking about my next venture; more than a dozen trips to Australia to launch a life-saving technology; and helping half a dozen other space entrepreneurs realize their dreams.

And then this: Discovering a resilience I did not know was in me and the humility that now informs my personality because of the near-loss of a company.

While I was leading Starsys, there was a Teddy Roosevelt quote that I thought of now and then but that made me squirm. Nothing at the time I would want on my office wall. The quote is from a speech Teddy gave at the Sorbonne in 1910:

> "It is not the critic who counts; not the man who points out how the strong man stumbles, or where the doer of deeds could have done them better. The credit belongs to the man who is actually in the arena, whose face is marred by dust and sweat and blood; who strives valiantly; who errs, who comes short again and again, because there is no effort without error and shortcoming; but who does actually strive to do the deeds; who knows great enthusiasms, the great devotions; who spends himself in a worthy cause; who at the best knows in the end the triumph of high achievement, and who at the worst, if he fails, at least fails while daring greatly, so that his place shall never be with those cold and timid souls who neither know victory nor defeat."

FROM THE GARAGE TO MARS

The focus of the quote ... *that it matters less whether you win or lose, but rather, how you try,* made me uncomfortable at the time. After all, I was succeeding. What would I be if Starsys was not successful under my leadership?

It was an amazing 20 years. An adventure-dream-come-true. What I thought was and would always be my career apogee turned out to be simply a chapter in a larger story. In retrospect, my odyssey with Starsys was exactly as it was meant to be. The perfect portion of my life, a great adventure, and the ideal transition to what was to come. The "man in the arena" quote now resonates with me. The Starsys journey was far from a failure. Instead, it created a family that continued forward doing great things. And, again, as the Starsys door closed for me, another opened to opportunities I couldn't imagine at the time.

There is no way I would press that button.

THE RED BUTTON

Starsys Research Corporation circa 1994

Starsys Research Corporation circa 1997

FROM THE GARAGE TO MARS

Starsys Research Corporation circa 2004

Buzz Aldrin (right) meeting with Starsys team, 2006

THE RED BUTTON

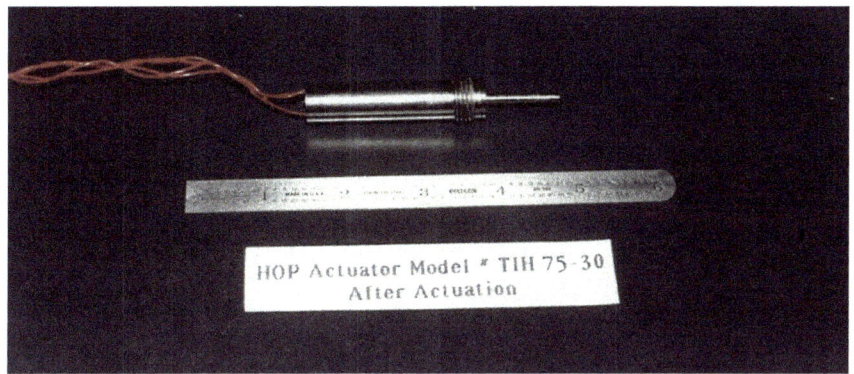

Starsys HOP Actuator circa 1989

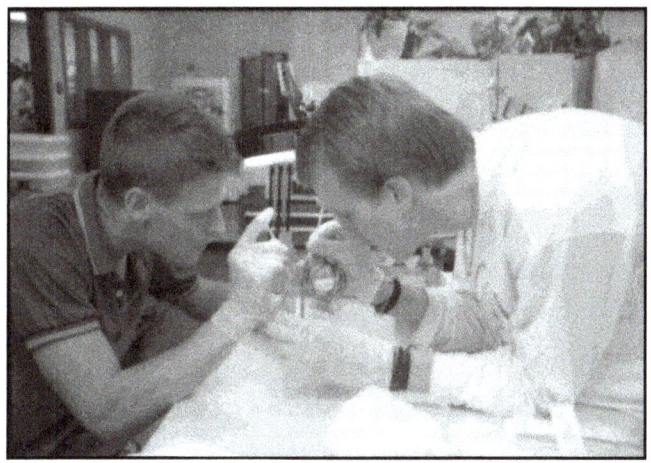

Assembling HOP actuators

Starsys HOP Actuators circa 1992

TRUE NORTH

> ... *The choice that must not be missed is to cherish your human connections, your relationships with family and friends. For several years you've had impressed upon you the importance to your career of dedication and hard work, and of course that's true. But as important as your obligations as a doctor, a lawyer, a business leader will be, you are a human being first and those human connections with spouses, with children, with friends are the most important investment you will ever make.*
>
> *At the end of your life, you will never regret not having passed one more test, winning one more verdict, or not closing one more deal. You will regret time not spent with a husband, a child, a friend, or a parent ...*
>
> —Barbara Bush, Wellesley College Commencement Address, June 1, 1990

WHILE LEADING STARSYS, A TUG of war had developed conflicting my purpose and identity. As time passed, I found myself beholden to two families: My first family at home—Jackie, Ryan, and Alyssa. And the company I had created and the people in it; 150 individuals with a shared mission I also was responsible for and cared deeply about. As the company grew and succeeded, I found my identity being defined less by who I was and more by what Starsys achieved. Over the 20 years, I had become my company.

In my mind, when people complimented the company, they were complimenting me. When the company achieved greatly, I did also. It was a subtle twisting of identity and a seductive affirmation of self as Starsys achieved. And as I handed over the keys to Starsys, I felt I was relinquishing my identify, my purpose, my meaning as I signed the acquisition documents.

I remember that disturbing dream from that time I wrote of earlier—the small model of the Starsys building in my arms, holding tight as if protecting my child, a faceless person reaching for it and forcefully pulling the building from my hands, me trying to hold on but failing, my identity and self, pulled from my grip. It was a metaphor of my experience. Something being extracted from me so intrinsic to who I was, I did not see how I would survive. I was laid bare; vulnerable, exposed, confused. It took a trip to Scotland with my son to shine a light on a lost truth.

* * * * *

At the time of the merger, I was approaching 50, and the three-year stint at SpaceDev was my chance to get all in order for the next 50 years. The slowing of my life after the merger was like a yellow-flag pit stop at the Indy 500.

I decided to get my body parts changed out and fixed up. Knees that had been trashed from 30 years of skiing were replaced with titanium and plastic. Shoulder rotator cuffs that had been shredded from decades of Ultimate Frisbee were sewn up. As one recovery was complete, I headed back to the hospital for the next. It was a *Macarena* surgery protocol, alternating from the right knee to the left shoulder, back to the left knee, and up again to the right shoulder.

Through all of this, I was trying to shed responsibility and it was hard. Although no one was pushing me for results, nor dependent on what I achieved, it was not a simple thing for me to relax and enjoy the earned peace. I was still going into work earnestly, one of the first to work in the morning, despite that what I was working on mattered little. One of my primary roles was simply to author press releases for the company.

I had often travelled as the face of Starsys, attending space workshops in the US and internationally and that continued with SpaceDev. A year after the merger, an opportunity arose to travel to Glasgow, Scotland, to an international aerospace conference that included a session on space entrepreneurship. I put together a paper that combined my story with that of two other successful space entrepreneurs I knew, comparing and contrasting our three stories. The paper was accepted and I made travel plans.

A couple of days later, I told my son, Ryan I was going to Scotland. He paused, weighing the odds of rejection, and decided to risk an ask; "Dad, could I come with you?"

I had travelled all over the world for business and was used to the question, my automatic response being "I would love that, Rye, but it just won't work this time, maybe next trip." My brain formed the response, and in the instant before I said, "Maybe next time," I paused as another Nudge materialized.

There was really nothing I had to do in Scotland. Even presenting the paper was inconsequential, something to fill my time at SpaceDev and provide the appearance of purpose.

"Rye, that might work this time. Let's see if we can pull it off."

We did. Ryan took time off from school, and I purchased an extra ticket. I explained that he might be bored stiff, hanging out at and aerospace symposium for a few days. He said that was just fine.

British Airlines had just opened a new direct route to Heathrow and was looking to impress new customers. We checked in for the flight, asking for an empty middle seat between us to make the overnight flight bearable. The gate attendant paused and smiled; "I can do better than that. We've got room in first. Would you like a complimentary upgrade?" We boarded the plane, took our seats, I accepted a glass of champagne, and we both grinned with delight.

We flew into London, taking a couple of days to do what you do in London—seeing shows in West End theatres, eating fish and chips, riding the London Eye, and taking the Tower of London tour. It was an enjoyable weekend simply because we were doing what you are supposed to do. After the long weekend, we took a train to Glasgow to the symposium.

Glasgow was wet, cool, and industrial gray. Ryan met dozens of our friends and business partners from around the world. He had known we had made space parts, but he discovered the reach of our company was broader than he had imagined. He attended the session on space entrepreneurship and watched my talk.

There was not much to do between symposium sessions until we found a virtual reality simulator that looked like a Starship Enterprise escape pod perched on hydraulic actuators, allowing it to violently tilt up and back and side to side. It provided the occupants a visceral experience of riding a roller-coaster or launching into space, including an overwhelming feeling of nausea at the end of the ride.

The latter didn't bother us as much as others at the conference as we repeatedly found ourselves the only passengers, giving us the opportunity to sit in the front row, fully immersed in the experience. Despite a lingering scent of Lysol blended with what it had cleaned up, when the attendant opened the door to let us out at the end of the ride, we would say "One more please!" He would shake his head, close the door, and send us on another ride.

After a couple days of wandering around various space displays, virtual roller-coaster rides, and collecting space swag, we both had symposium cabin fever. We were to attend for a few more days, but the Nudge objected.

"Rye, there really is no reason for us to stick around. Let's get out of here."

We rented a car, and blowing off any remaining responsibility, drove north, stopped to skip stones on Loch Lomond humming "You take the high road and I'll take the low road..." with a new appreciation of the lyrics.

We headed towards the castles, lochs, and the highlands, driving at high speed on the wrong side of the road, both of us keenly focused. Each highway roundabout we approached triggered a strained, urgent voice from Ryan, "Left side, Dad! Stay on the LEFT side. Could I PLEASE drive???"

My son found it hilarious that every time we were at an intersection, I would turn on the wipers instead of the turn signal, a habit almost impossible to break with the reversed controls on the steering wheel. Through force of will, I tried to use the correct lever but continued to fail, each instance triggering hilarity from the left seat and "Son of a bitch! I did it again???" from the right.

We travelled for three days without schedule, waking, and getting in the car, picking a destination and just going. The singular agenda we shared was to visit Loch Ness and take a swim just because we could. When we arrived, the loch was disappointingly similar to every other loch in Scotland. Brain-freeze, cold, dark water tucked within the shoulders of old mountains. We drove to the first access point we could find, a beach of small stones between road and water. We both wanted to get a picture of us swimming in the Loch. I was committed to skinny-dipping. Ryan was mortified.

I set up the camera, Ryan waded into the dark water, going far enough from shore to squat down so that it looked that we were swimming, but close enough that I could join him after pressing the timer on the camera. I stripped off my clothes. Ryan hung his head in embarrassment. I pressed the shutter and ouched and yelped my way across the sharp beach stones into the water towards Ryan, and "click" ... the camera took a picture of my bare, white behind and my son's ashamed expression. Back to the beach, I pressed the shutter again, ignored the stones, hustled out into the water, and got the shot.

As we were driving from here to there, we glimpsed a deserted castle, stranded at the end of a barely discernable dirt road. A quick U-turn, a hunt for a break in a stone fence, and a drive down the derelict road led to a dead-ringer for a castle from a Monty Python movie. We spent an hour wandering the 800-year-old building, there for us alone to explore.

Driving from one side of the country to the other through the highlands was beautifully desolate. We didn't talk much, just taking in the treeless, mountainous landscape. With nowhere to be and no when to be there by, options opened up. At one point, we passed a small mountain and looked at each other and said, "What do you say we climb a Scottish mountain?" We stopped and hiked up the wet, spongy trail, so different than the dry, hard trails of Colorado. At the top, we set the camera up and took a picture. Nothing overly spectacular in the background, simply a father and son memorializing a hike.

The trip continued to the East Coast and then south for an evening in a small city called Killicrankie, a town tucked next to a verdant gorge that we rode rented bikes through, christening it Narnia for the magical feel of the place. We finished the trip in Edinburgh. Our stay there was brief. A single day, a quick ghost tour of the catacombs under the Edinburgh streets, and then we

Father and son in Scotland

were headed to the train station for the overnight trip back to London.

We arrived a couple of hours early and headed across the street to the Balmoral Hotel, where they served us high tea in a walnut-paneled, hundreds-of-years-old, sitting room. We sat in overstuffed chairs and sipped Oolong with milk and sugar, sure we were sitting in chairs that had held kings, queens, lords, and ladies.

Leaving the hotel, we met a couple of pretty twenty-something girls in evening gowns taking a break from a reunion. We talked for a bit, and when they realized it was Ryan's first trip to Scotland, one leaned over planting a kiss on his cheek. "Now you've had your first kiss from a Scottish lass!" Ryan blushed. I grinned. We crossed the street and boarded the train to London for our flight home.

I so hoped the trip had been something special, a trip of a lifetime for both of us. On the flight back, as I reflected on the trip, I began to question if it had been as extraordinary as I had thought. I began to focus on what we had missed. As we were flying over Iceland, I finally said "Rye, I sure wish we had had more time in Edinburgh. It seemed we missed out by spending only that one day".

He interrupted me, "Dad, the trip was perfect. I wouldn't change a thing".

"Huh? Really? What was your favorite part, seeing *Stomp* in London? The castle outside of Glasgow?".

"No, it was driving through the Highlands, just me and you, no need to talk, no agenda, just being together. Being able to say "Hey, let's climb that mountain." and then doing it.

*****

In 1990, Wellesley Women's College had been looking for a commencement speaker. The committee settled on former First Lady Barbara Bush; their first choice, Whoopi Goldberg, had been unavailable. There was consternation on campus that a grandma and stay-at-home mom was speaking to thousands of women who were preparing to change the world on graduation. Mrs. Bush recognized that she was not the student body's first choice. Within that context, she gave a speech that is considered one of the top 50 speeches of the 20th century. A quote from that speech introduces this chapter and is on a plaque that has been on the wall of my office for dozens of years.

Years from now, when my children are asked "Tell me about your dad," I pray the response is not "We didn't get much time together," "He worked weekends," or "We didn't see him much; he had to save his company."

I hope for "He's the guy who showed me how to fish," "He always helped me with my homework." "He taught me how to swing dance." "I could tell him 'There is this guy in school who likes me!" and he would listen carefully and offer advice."

But as Starsys went through the rocket-sled-ride of explosive growth and its near-death experience from gorging on its success, I felt that I was so scrambled from the tempest of the past years that I had lost the compass heading to that which was truly important to me. That I had gambled all on the company and lost.

At the end of *The Wizard of Oz,* there is a lesson the wizard conveys to Dorothy and her group: "You always have had the ability to go home. Just click your heels. The things each of you thought you had lost were with you all along—your home, your heart, your brain, your courage. I simply gave you something that reminded you of what you had forgotten".

A five-minute read of a journal entry shortly after we returned from Scotland was the three ruby-slipper taps that brought me home. A journal that had come to be more than a decade before.

* * *

I had travelled quite a bit for Starsys and frequently found myself on planes crossing the country, missing my family. When Ryan was five and Alyssa one, I was feeling particularly distant from family. I opened my laptop and began two letters, one to Ryan and one to Alyssa. They both started similarly:

> *"Dear Alyssa,*
> *It has been a couple of months and a year since you joined our family, and I wanted to tell you a bit about you ... you wake me up every morning at 6 and while everyone else is sleeping, we go outside, you sit on my lap and we watch the day wake up ... "*

> *"Dear Ryan,*
> *You are growing so fast. I worry that I won't be able to call you sweet pea much longer ... you are 5 now, and growing up ... you care so much about doing things the right way. Mom and I can see it in everything you do ..."*

It was an easy thing to do, and it became part of my travelling routine. The first several letters were simply that ... a quick reach-out to Ryan and Alyssa, creating a connection of sorts despite being away. But at some point, I realized the value was not in each letter itself, but that I was creating an archive of the journey as Jackie and I developed as parents and Ryan and Alyssa developed into young adults. I continued writing the letters every couple of months, year after year as they grew into pre-school, elementary school, middle school, and then high school. As the compilation of letters became 10, then 20, then 30 pages, I recognized they had become something precious and special. By the time Ryan was 15 years old, there were 40 pages. A record of our relationship and the our growth as he went from kindergartner to travelling partner.

The trip to Scotland was a transition point for the both of us, and I recognized it as the opportunity to present the letters to him. I went to a stationery store in Boulder and found a journal made from

FROM THE GARAGE TO MARS

> **Tibbitts Tip**
> **Journal Your Parent Journey**
>
> The big things in our life often come from a multitude of small bits and pieces that build to something extraordinary simply through discipline.
>
> The letters to Alyssa and Ryan didn't take much work at the time, 30 minutes here and there while relaxing on an airplane. I had no idea at the time of the profound result 15 years of those moments would create.
>
> The precious records created of our parent-child relationships as they grew from dependence to sending them out in the world.
>
> I've told the story often to young parents, suggesting they consider a similar journal for themselves.

distressed leather. I printed out the story in a calligraphy font on parchment paper, using bookbinders' glue to attach each letter to a page in the journal. The result was appropriately wrinkly and weathered.

In the first pages of the journal, I wrote about how the letters had come to be, that it was a record of how we grew as father and son. I wrote of how I hoped that although the journal might not be something that he frequently read, that it would be a record that he would come back to as he became a father, and lastly how proud I was of him, and what a wonderful journey it had been—and was—being his dad. I closed with talking about how I enjoyed giving him the occasional backrub to put him to sleep through the years, a way for me to show my love, but that I expected those times would soon come to a close if they had not already.

I had left dozens of blank pages in the back of the journal and on the first of these, I wrote about our time in Scotland, the things we had done, his and my favorite memories, and suggested that the remaining pages might be a way for he to record future adventures and stories.

The love between fathers and sons is often covert; a quiet thing, told by the occasional hug, a shared smile, a comfort together that doesn't necessarily require talk. Companionship that requires little

TRUE NORTH

ceremony but just is. Honoring this context, I simply left the book on his bed a couple of days after the trip, not saying anything, but thinking that he would find and read the journal later that day.

Ryan said nothing of it over the next day. I wondered if he had found it but had not taken the time to read, or possibly that it had meant less to him that it had to me.

Later in the next day, I had reason to look at the journal again; a detail I wanted to confirm. He was out of the house, and I went to his room, the book on the nightstand. I sat on his bed and picked it up. As I thumbed through, it opened to a few freshly written pages. He had already added to the journal, his writing following my summary of our Scotland trip.

The words began: *"Dad, I don't know if you will ever read this, but if you do, I want you to know how thankful I am that you are my father...."*. He went on to write about what he had appreciated most and remembered best of our time and adventures together. He closed with a simple statement that was a call-back to my journal preface: "Dad, and there will always be time for backrubs...".

FROM THE GARAGE TO MARS

Like a vague dream suddenly remembered in its entirety, the letter forced me to recognize that through it all I had stayed true to what was truly important. I had been the dad who missed work to chaperone school field trips. I had worked hard to not work weekends. I had danced with wild abandon with my daughter when no one was looking and taught her to swing dance for when people would be watching. I shared with her that the chords C, Am, F, and G were the keys to the kingdom of playing pretty much every ballad ever written by ear, years later lying on the couch with Milo the Wonder Dog at my feet, listening in awe to the beauty of her piano playing. I had been (and 20 years later, still am) the official storyteller at Niwot Elementary School .

My arms had wrapped around Ryan as he etched his name on hardware headed to Mars. I had spent endless nights relearning math to help with homework. More than once I had leaned against my daughter as we watched father-daughter movies, her admonishing me with slight smile, "Dad, you aren't going to cry again, are you?" her, of course, hoping that I would. I had missed key company meetings to not miss dance performances. I had coached Ryan's baseball team despite adding little value and led renegade boy-scout trips to the top of mountains, down rivers, and into caves.

Ryan and I had built dry-ice cannons in the backyard that sent golf balls 500 feet in the air the two of us scurrying to make sure we didn't end up with divots in our skulls. And finding closure on a prior uncompleted project with my dad, Ryan and I delivered a working hovercraft for a 4th-grade school project, sending his classmates one by one floating and spinning across the gym floor.

Five years after selling Starsys, my daughter Alyssa turned sixteen, and the journal I had written for her chronicled fifteen of those sixteen years, stories of triumphs and tears. Her looking at me while we were driving somewhere and saying "Dad…there is this boy I really like, and I think he likes me!" My heartbreak when she fell and broke her wrist as I pushed her to try roller blades too soon. Alyssa waking me up by dropping a Barney doll containing two D-batteries on my head when I fell asleep next to her crib.

When she was sixteen, I surprised her with her father-daughter coming-of-age trip. A week in Australia where I was spending time deep in my next venture. As with Ryan, after we got home I presented her with my record of our growing together. And then, five years later, deep into COVID, she gave it back to me, with 20 additional pages continuing the story as she graduated high school and went on to study neuroscience at Colorado State University.

* * * * *

Starsys had been an amazing journey. A space company created from a tube of wax. Our logo on Mars. Floating weightless in the Vomit Comet. Initials circling Saturn. Being a part of mind-blowing discoveries on Mars, Saturn, Mercury, and Pluto. But that was simply icing on the cake.

I closed the journal, tears welling, reminded, and reaffirmed that although my self-worth had been battered, I had not wavered from what was truly important to me. I had never lost my True North. That human connection trumps all.

Late in life, as I reflect, I will not be thinking of the first picture back from Mars or Dreamchaser. I will be remembering a moment on a bed reading my son's words, and other moments like it, and not regretting time not spent with a wife, a child, a friend, or a parent. I wish the same for you.

EPILOGUE

It was early afternoon, May 8th, 2008. I was driving to meet Dave Sueper, Vice President of LGS, a small Telco in Thornton, Colorado. Two years prior I had signed the documents that had handed over the company I had created, Starsys Research, to SpaceDev, and I was looking for my next venture.

Over the past two years, in partnership with Mark Sirangelo of SpaceDev, the University of Colorado, and $2M of funding from Congress, I had established eSpace, the Center for Space Entrepreneurship. eSpace was the first congressionally funded space company incubator in the world. We provided $20,000, office space, and mentorship to promising, young space companies in trade for 5% in stock. It had been wildly successful; by sharing lessons learned from creating Starsys, we were able to help launch a half dozen space companies, one of which, Blue Canyon Technologies, sold for $350 million dollars 15 years later.

Ripples from Starsys' heretical approach to making space hardware by emphasizing Fun and Family spread, changing how Colorado space companies did the space business. A decade later, I would visit thriving space companies, see CEOs dressed as Elvis singing "Hunk a hunk of burning love" at the company Christmas party, or hear about companies investing heavily in helping disabled veterans, and smile, recognizing we had made a difference in more than our company.

But mentoring emerging space companies wasn't filling my cup. I found out I was a serial entrepreneur and hungered to found another company. To guide growth as CEO, with success or failure a result of my distinctions and decisions.

I had several irons in the fire at the time: Syberenity, an app to help alcoholics stay sober; NickelClick; an app for children to contribute to good causes a nickel at a time; and LockOut, a clever contraption to prevent theft from commercial vehicles. They were all interesting. None were compelling.

I came to the intersection near the LGS building. As I turned left, I noticed the asphalt scattered with the shattered glass of a recent accident. It didn't look serious, probably just a fender bender. I pulled into the parking lot, walked in the door to the reception desk, and asked for Dave. The receptionist eyes were red, her face ashen. She teared up. "I'm sorry, you can't meet with Dave today. He was killed an hour and a half ago in an accident in front of our building."

He had been coming back from lunch. As he drove through the intersection, a 16-year-old teen had driven through a red light, T-boning Dave's car on the driver's side, killing him instantly. The teen had been T9* texting. The police estimated he had been looking down at his phone for more than 6 seconds before he hit Dave at 35 mph.

Dave was similar to me in age and a husband and father of two, as I was. If I had been an hour and a half earlier, that could have been me. I drove home shaken.

A question formed that wouldn't let go. Could I do something about it? Texting while driving was becoming a social crisis, and it was going to become much worse. The iPhone had been released a year earlier; phone distractions were getting much harder to resist. In addition to most adults, most teens of driving age had cell

---

* T9 texting was how messages were typed on flip phones. Any of 26 letters could be typed by hitting one of the nine number keys, multiple times.

EPILOGUE

phones. The texting demographic was changing. Distracted drivers would soon include parents, commuters, soccer moms.

It was clear the solution would have to come from technology; human nature meant that we all would be saying to ourselves: "Just this one won't hurt anyone. I'll be careful", but like an open bag of Lay's potato chips, once you eat one, you eat another and another until the bag is empty.

The obvious solution—"Just put an app on the phone"—was likely to fail, as an immutable rule of app design is that one app cannot affect another. For instance, WhatsApp can't affect Instagram and vice versa.

Entrepreneurial ventures usually come to be as a solution to a Big Problem, and this was already an enormous problem, soon to be much worse. The question of "Can I help?" morphed into "What will be the ultimate solution, and could I could help create the invention that provided that?" I decided to bring together a group of smart, successful technical entrepreneurs to brainstorm a provocative solution, launch a company to bring it forward in time, and possibly save thousands of lives.

I brought together Merc Mercure, co-founder of Ball Aerospace, Margert Burd, founder of Magpie Telecom Insiders, and Frank Tai, founder of Technology Advancements Incorporated. We considered every possible angle and arrived at what we believed, and ultimately became widely regarded as THE solution:

Stop the distractions at the network level, before they get to the phone.
No app is required.

With this approach, no app on the phone is required. Instead, put a small device in the car that is connected to the cloud that alerts software.

With a typical car having no more than three potential drivers, use AI to determine which of the possible drivers is actually driving from patterns in their driving behavior such as: what speed do they typically go on a given road, do they roll through stop signs or come to a full stop, where the drive starts from—a school, a particular business, a store that a driver frequents.

- Alert the Telco to use their existing content filters to block distracting content of the driver's phone while continuing to allow features such as navigation and music streaming.
- Gamify the system. Keep track of streaks of non-distracted driving and provide scores and badges for the safest drivers. Provide leader boards to allow competitions within a family, or within a high school.
- Incentivize the game by awarding safe drivers free Chipotle burritos, discounts at Starbucks, or twenty cents per gallon discounts.

Or, effective within the Gen Z demographic, pay it forward by giving teens an option. "Would you like to redeem your award for a burrito, or feed a family of four in Ghana for a week?"

- Provide the information to insurance companies so that they can provide 20% discounts to the safest drivers.
- Create a technology that would be paid for by saved lives and fewer accidents.

Focus groups indicated that 53% of teens would ask their parents for a device that rewarded them with a couple of burritos a month for their non-distracted driving—and tell their friends about it.

 I had found the Big Problem to commit to. The company we formed was called Katasi, a derivative of the Greek word for "hush." The technology was called Groove. The mission: to make distracted driving an unacceptable human behavior.

EPILOGUE

We introduced the solution to insurance companies, congresspeople, legislators, press icons, commercial trucking companies, state patrol organizations, and others. The result became predictable—polite skepticism, followed by the moment they grasped how it worked. Then an epiphany and some form of the statement, "That's it. That's the solution!"

Groove could ultimately save tens of thousands of lives. It could eliminate thousands of late-night visits by a policeman to a parent's home that would start with, "I'm so sorry, but there has been an accident…" And the lives saved would never know it would have been different without Groove.

I fully committed to bring Groove to life. It became a more-than-12-year journey with extraordinary experiences, mind-bending highs, and soul-crushing lows.

- American Family Insurance, hearing about our technology, urging us to come to their headquarters in Madison immediately to discuss, and shortly after investing a million dollars and deploying their gravitas, encouraging first Verizon, then Sprint, then T-Mobile to deploy Groove.

- My being the first to drive with the new technology and being the first to experience a network-enabled distracted driving solution.

- Cold-calling Matt Richtel, Pulitzer-prize winning journalist from the *New York Times*, who patiently took my call, but then checked with industry experts, calling back two days later saying he got his editor's OK and was flying out the following week. His embedding with us for a couple days, experiencing the technology and attending focus groups, then writing a front-page article in the NYT business section with a half-page picture of me and my son above the fold.

FROM THE GARAGE TO MARS

- Broadcaster Katie Couric reading the article the Sunday after. My picking up a call in our office later that week, thinking it was a toner salesman, and instead, it was Katie's producer. The producer then flying the widowed wife of Dave Sueper and me to New York for an interview and a drive around Manhattan, showing Katie how Groove works. The national broadcast of the eight-minute piece crashing our servers as 120,000 viewers tried to buy Groove the next day.

- Receiving a call from a business icon in Australia saying "Mate, how about coming down here to raise investment and deploy Groove?" It taking him three calls before I realized he was bona fide. Flying to Sydney, catalyzing four million dollars in investment and deployment with the largest Telco in Australia.

- High-profile segments on CNN, *The Today Show,* and *Good Morning America,* giving us national attention and leading to a reach-out from a mother in Kansas saying, "I've been silent for years, but your segment on CNN had me realize it's time to help." Two weeks later, receiving a raw six-page chronology of ten days, six years prior, describing how her daughter had been killed in a distracted-driving accident on a Friday and then seven days later, her son killed the same way. The letter closing with the statement: "I've been silent long enough. How can I help?"

- Every major Telco in the US partnering with us, and then pumping the brakes from concerns about liability of it working.

- Meeting with a powerful legislator from a West Coast state in a walnut-paneled room in the state capitol, his commitment to making Groove mandatory in his state sealed with a firm handshake, a pull in for a hug, and whispered "Let's f...'ing do this."

EPILOGUE

- A month later, the mother who had lost two children, agreeing to fly out for a national press conference to support the legislation. Five days before the press release, the legislator being corralled by lobbyists from a major Telco, and his call shortly after: "Not only will I no longer support this bill, I'll kill it if it's moved forward." Calling the mother the next day and telling her the press conference was off.

- A decade later, the innovations we developed deployed by insurance companies to more than 20 million drivers. A story to be told in a following memoir.

* * * * *

In its simplest form, entrepreneurship means creating something from nothing. Tesla, Bezos, Edison, Jobs, DaVinci, Musk have impacted the world by imagining, and then making it so.

    I had no idea that seven dollars in hardware-store parts, some wax, and a certainty that, "This is so cool. There has to be some use for it..." would lead to my three-decade journey, which is far from over. I'll keep you posted.

ACKNOWLEDGMENTS

Alison Mahan Tibbitts—Your dream of writing and publishing your story never got its chance but sparked a son's dream of writing and publishing his story. This is for you, Mom.

Theodore William Tibbitts—You are my hero, my inspiration, and the man who taught me that it is OK to unbuckle the seatbelt now and then to see what is around the corner, under the rock, or in the galley of a 747 during heavy turbulence.

Tia—Your loving support and wisdom have always been there for me when I needed it most, whether I knew it or not.

Jackie, Ryan, and Alyssa—You lived the journey with and beside me, celebrating the triumphs, weathering the rough sailing, and throughout believing in and supporting the dream.

Scott Christiansen—None of this journey would have happened if you hadn't believed in what I was doing, thrown in with a "one person and a garage" space company, and made our shared vision come to life. Thank God for Boulder Falls.

Mark Sirangelo—You were the visionary and catalyst who brought our companies together, then took the helm of the Starsys team and went on to do a thing amazing and unimagined: Dreamchaser, changing how humanity will reach the stars.

FROM THE GARAGE TO MARS

And Kira Henschel—Meeting and working with you has been a blessing and a joy. You saw what FTGTM could be, and together we made it so. And to Jeanette, who recognized what Kira could bring to the project and became the matchmaker who brought us together.

And finally to the entire Starsys family—We together created something extraordinary that made a difference in the world, and we had a hell of a good time doing it.

Thank You!

ABOUT THE AUTHOR

Scott Tibbitts was the founder and CEO of Starsys Research Corporation. From its humble start, Starsys grew to become became the world's leading supplier of mechanical devices for spacecraft. One key accomplishment was delivering more than 27 actuators enabling the Mars rovers *Spirit* and *Opportunity* to explore the Martian surface for more than 25 times the expected mission lifetime. Starsys became known in the space industry for its legendary corporate culture that delivered extraordinary results by inculcating *Fun* and *Family* deep into its corporate DNA.

Scott lives in Boulder, Colorado, where he continues to support the aerospace entrepreneurial ecosystem in Colorado, California, and Australia, providing mentorship and coaching for space entrepreneurs, and helping organizations transform their culture through Fun and Family.

He is an internationally sought-after keynote speaker on the subjects of space, entrepreneurship, and the creation of legendary corporate cultures. He is actively involved in Colorado universities as a guest lecturer for aerospace and entrepreneurial programs, and is the international aerospace mentor for the University of Southern Australia ICC Incubator.

FROM THE GARAGE TO MARS

He is also the official storyteller for Niwot Elementary School, and the keyboard player in the classic rock garage band "Too Much Fun," renowned for playing the same 40 songs in the same order for more than 25 years.

Scott is available for select engagements for

Keynote speaking
CEO Mentorship
Leadership training
Organizational culture transformation

To connect with Scott, contact with him via his website:
www.scotttibbitts.com

Or email at
info@level6.space

Or simply scan the QR code below.

www.ingramcontent.com/pod-product-compliance
Lightning Source LLC
Chambersburg PA
CBHW071707160426
43195CB00012B/1603